Zero Mass Design

A Strategy for Designing a Better Future

David D. Thornburg, PhD

Constructing Modern Knowlege Press

Zero Mass Design: A Strategy for Designing a Better Future

Published by Constructing Modern Knowledge Press, Torrance, CA USA

cmkpress.com

ISBN

Paperback: 978-1-955604-08-6

Hardcover: 978-1-955604-09-3

DES011000   DESIGN / Product

EDU008000   EDUCATION / Decision-Making & Problem Solving

EDU029030   EDUCATION / Teaching / Subjects / Science & Technology

TEC062000   TECHNOLOGY & ENGINEERING / Project Management

Layout: Sylvia Martinez

Cover: Yvonne Martinez

"*The excellence or beauty or truth of every structure, animate or inanimate, and of every action of man, is relative to the use for which nature or the artist has intended them.*"
Socrates

"*Anything forced is not beautiful.*"
Xenophon

"*Form ever follows function.*"
Louis Sullivan

"*World revolution is upon us; Humanity has two options:*

1. a bloody, pull-the-top-down, political revolution, or

2. a bloodless design science revolution of exponentially compounding 20th century science and technology gains in accomplishing both ever greater and more incisive tasks with ever less resources per each accomplished function for ever more humans."
R. Buckminster Fuller

"*The best way to predict the future is to invent it.*"
Alan Kay

"*We're all designers now; so we may as well get good at it.*"
Chris Anderson

Acknowledgements and Dedication

An early draft of this book was shared with my graduate students in the Design Division at Stanford University many decades ago. I thank my students for their honest assessments of my ideas and especially thank Steve Paley for sending me a copy of the original manuscript when my only copy was lost in a flood. Steve's book (Paley, 2010) on the art of invention is a great read on its own, and is highly recommended

This book is dedicated to all my friends, colleagues and students who have used the ideas in this book to invent the future.

"I have no doubt that it is possible to give a new direction to technological development, a direction that shall lead it back to the real needs of man, and that also means: to the actual size of man. Man is small, and, therefore, small is beautiful."

E, F. Schumacher

Contents

Foreword

In 1983, Professor David Thornburg handed out an unfinished manuscript titled Zero Mass Design to a graduate-level class in the philosophy of design that he was teaching at Stanford University. He had formulated a novel approach to design thinking, and he wanted to share his ideas with his students. I was at the time pursuing a master's degree in Stanford's Product Design program, and the alluring title suggested an innovative and perhaps totally different approach to engineering design. I eagerly read the manuscript and was not disappointed. David proposed a concept that was so original and profound from a design standpoint that it stayed with me as a guidepost throughout my professional career in industry and technology.

Often, when we are solving a problem or designing a product, our instinct is to pursue greater functionality by making our solution more complex. The more functionality we want, the more complexity we feel we must add. But David's very original idea is that we can gain functionality in another way. That is, by driving simplicity into our solution. As we attempt to solve a problem—this applies to problems both within the technology sphere and without—we should challenge ourselves to find the simplest possible solution. And then make it even simpler.

What David proposes is that as we pursue simplicity, we will also find that functionality also increases. He illustrates this with a U-shaped curve with increasing functionality on one axis and increasing complexity on the other. He calls the drive toward improving functionality by making a solution more complex, engineering; and the drive toward creating more functionality by making a solution simpler, invention. Engineering is a straightforward process; invention is harder and requires some of the techniques of Zero Mass Design.

I believe that the concept of Zero Mass Design should be taught in every engineering school as one of the fundamental approaches to problem solving. David has now updated his former manuscript into a full-length book where he expands upon many of his original ideas. This book not only explains the concept of Zero Mass Design in detail, but also provides real-world examples of its practical use, many from the author's own inventive work. It has been written to address makers in all communities, including students in elementary and high schools, as well as the original audience of college students exploring the principles of design.

I commend this important book as an original and powerful approach to problem solving. I hope that it will have as much influence on you as it had on me.

Steven Paley

Author, The Art of Invention

Preface

The ideas behind this book were triggered by a concept for corporate budgeting developed in the 1970's, zero-based budgeting (ZBB) is a method of budgeting in which all expenses must be justified and approved for each new period. Developed by Peter Pyhrr in the 1970s, zero-based budgeting starts from a "zero base" at the beginning of every budget period, analyzing needs and costs of every function within an organization and allocating funds accordingly, regardless of how much money has previously been budgeted to any given line item. Depending on a corporation's growth plans, this budget might be larger or smaller over time.

Based on these ideas, I decided to apply a similar concept to the process of designing objects, and coined the phrase Zero Mass Design (ZMD) when teaching a graduate course on creativity as an adjunct member of the design division faculty at Stanford University. ZMD is a methodology based on the idea of first identifying the core function of an object under consideration, and then finding the minimum way of achieving this goal. This differs from an incremental design process based on the idea that a new design might come from (for example) an attempt to reduce the cost of an existing device. The results can be striking. For example, a goal might be to reduce the cost by 10%. Using the principles of ZMD, it is not uncommon to find that the new design has a lower cost by a factor of ten!

In this book, we'll explore the process of ZMD through various examples, spanning a wide range of fields. While some of the examples may seem frivolous, they are not. In fact, at least two corporations have been started by applying the principles I'll share, and the reader is likely to develop inventions of her own that are triggered by the ideas explored here.

In February, 2005, Dale Dougherty published his first copy of MAKE

Magazine in which he said, "More than mere consumers of technology, we are makers, adapting technology to our needs and integrating it into our lives. Some of us are born makers and others, like me, become makers almost without realizing it" (McCracken, 2015).

The Maker Movement uses modern tools like 3D printers, laser etching, and electronic gadgets like the Arduino, and opened the door for just about anyone to make the transition from consumers to makers.

In fact, while this book started out as part of a university course, the ideas incorporated here apply in any place where creative design takes place—schools at most grade levels, for example.

There are limits to the kinds of things we'll explore. In my view, there are three maker domains:

1. Recipe following
2. Repurposing existing artifacts
3. Inventive designs

This book does not deal with items made by following recipes. It does, however, apply to repurposing existing objects (sometimes called upcycling) and, especially to the domain of inventive designs. Our chosen domains fall within the scope of what the pioneering educator, Seymour Papert, called constructionism (Papert & Harel, 1991).

As Chris Anderson said: "We are all designers now, so we may as well get good at it" (2014, p. 53). And that brings us to this book about design. To be more specific, this book deals with design as a process involving human intervention and resulting in an artifact. The artifacts which result from this process, be they plans or finished objects, have certain characteristics which reflect the underlying activities which resulted in their creation. On a broad scale we can think of these characteristics as including functionality, beauty, simplicity, complexity, or numerous other subjective criteria. While this book grew out of a course I taught to graduate students in Design, it applies to professionals and itinerant tinkerers alike. No previous training as a designer is required. The ideas are applicable to makers of almost all ages and disciplines.

It is common for designers to be admonished to keep their work simple and functional. But while it is one thing to want to create a simple design, it is yet another thing to know how to go about doing it. The function of this book is not to catalog myriad examples of beautiful designs with the hope that these examples alone will somehow construct a useful philosophical framework in your mind. Instead, the object of this book is to explore the implications of extreme simplicity and to provide a structural framework or discipline in which simple design can take place.

Zero Mass Design is not a panacea, nor can it be used in all fields for all projects. Nonetheless, certain insights which form parts of this philosophy may be of utility to you, regardless of your personal style as a designer.

To me, the design process embodies two activities—creativity and discipline. Both are needed in order for the design process to take place. Neither is sufficient alone. This close coupling of creativity and discipline is found in works of music, silverware, a box kite, computer programs, paintings, telephones—it is present in designs from every form of art and technology known to mankind, and it is a coupled relationship which has existed from the moment man first shaped a stone or made marks on a cave wall.

Design is an activity which transcends occupational boundaries. Any activity in which something is intentionally crafted involves design. Because the author of this book is a technologist, the examples used to illustrate various concepts will be drawn largely from technology; but it is hoped that the principles described here will have utility to those who design in other areas, be they artists, computer programmers, weavers, or practitioners of the myriad other arts and disciplines.

This book is divided into two parts. The first part deals with the concept and mechanics of Zero Mass Design. Once this section is finished, you will have learned how this design philosophy is used and how it works. The second part acts as a bridge to explore some of the consequences of simplicity on the design of various artifacts, some of which have been in use for decades.

This book was written for designers and inventors in all disciplines. Great inventions come from many sources. Your education level, field of interest, and prior experience have little or nothing to do with the quality of your inventions. Mindset, on the other hand, can have a huge impact on your inventive ability. Personally, I have invented things in fields I never studied in school.

Because of my own experiences, I decided to write a book describing a particular inventive mindset—Zero Mass Design. While the principles of this design methodology have been applied by many inventors around the world, I came to these ideas on my own, and have applied them throughout my professional life.

My goal is that you'll find these ideas appealing enough to apply them to your own work.

An Invitation to K-12 Teachers

If you are a K-12 teacher, you might be wondering why you're holding a book about design principles. After all, the subject, by itself, doesn't seem at all connected to the existing curriculum.

In fact, the subject cuts across the curriculum and many grade levels. It can be argued that students are natural-born designers. Youngsters play with blocks and other construction toys, building elaborate structures from simple shapes. And a large empty cardboard box can become a spaceship or anything else that captures a kid's imagination. The role of the child as designer is explored in some depth by Alexandra Lange (2020), a topic explored almost a century ago by Rugg and Shumaker (1928).

The problem is that too many children are weaned from these ideas with a one-size-fits-all curriculum. Some schools have tried to address this topic with so-called "makerspaces," special rooms where kids have at least some freedom to make things of their own design. While these spaces are a step in the right direction, they perpetuate the idea that schooling should be focused on STEM subjects (science, technology, engineering, and mathematics) with little (if any) focus on the arts. In my view, this is a tragic mistake.

But imagine a school where students have the freedom to work on curricular topics with the kind of freedom associated with Makerspaces, no matter if the topic is mathematics or Macbeth.

This topic goes back at least a hundred years with the creation of the Bauhaus in Weimar, Germany, when gifted architects and other artisans did transformative work. During the second world war, the founders of this movement escaped Nazi Germany and one of the original team, László Moholy-Nagy, opened the New Bauhaus in Chicago. Another facility was created at Black Mountain College in Black Mountain, NC. While this college did not last a long time, it had some amazing people there including Buckminster Fuller, and John Cage, who worked alongside the Bauhaus founders Joseph and Anni Albers, with connections to the New Bauhaus group in Chicago. The Chicago facility is still operating today as the Institute of Design at the Illinois Institute of Technology.

One of the historical projects there was to sculpt a block of wood into a shape you'd be "happy to hold." The most surprising result of that task may be the bar of Dove soap, designed in 1952 by three students, Donald Dimmitt, William J. LaVier, and James Logan, as part of a special project funded by Lever Brothers. Their design is still used today. This was in keeping with the original Bauhaus' slogan, "Art into Industry," which spoke to their hope of bringing art into conversation with the systems of mass production.

My Mother was part of the Chicago school in the late 1930's and always told me that the focus was on art along with utility, probably because the original founders were architects. But this convergence transcended buildings, and was (and is) applied to all manner of artifacts, including painting, typography, architecture, textile design, furniture-making, theater design, stained glass, woodworking, and metalworking. Today we can safely add computer software to the mix. This implies that the arts and engineering, for example, are best explored in a combined manner.

This is true in most fields of study.

For example, when children study math they are generally taught math principles stripped of their beauty. Students focus on the mechanics of calculation with little or no exposure to the kinds of thinking they need to dig deeper into the kinds of creativity explored by professional mathematicians.

For example, children are taught about the Fibonacci series where the next number is the sum of the previous two. The first few elements are 1, 1, 2, 3, 5, 8, 13, 21 … The story behind these numbers is interesting, but to me the most fascinating part is that the ratio of successive Fibonacci numbers approaches the Golden Mean (1.618…), a number known to the ancients and found in the dimensions of ancient buildings, as well as the geometry of seed clusters in (for example) sunflowers. Fibonacci himself never saw the connection, and the proof of this connection came from Kepler, four hundred years later.

It makes one ask what other beautiful connections are awaiting discovery by children whose creativity and inquisitiveness could be fostered by inspired teachers like you operating in a school where transdisciplinarity is the norm.

In 2014, I wrote a book about the design of schools, *From the Campfire to the Holodeck: Creating Engaging and Powerful 21st Century Learning Environments*. My premise was that there exist four primordial learning spaces: the Campfire, Watering Hole, Cave, and Life. The Campfire is home to didactic presentations while the Watering Hole is home to Vygotsky's social constructivism. The Cave is home to Piaget's cognitive constructivism, and Life is the applied space associated with Papert's ideas about constructionism. My point was (and remains) that no one learning space meets the needs of children and that different kids learn the most when they can choose their own learning spaces as they need them—moving between spaces during the day as they need.

A consequence of this is that students in this kind of environment become versed in the concept of Epistemic Frames, a formulation explored by Professor David Shaffer (2008) in which there are five elements to any practitioner of a discipline: Skills, Knowledge, Identity, Values, and Epistemology. Traditional educational environments focus on the skills and knowledge of a field, with little if any focus on identity, values, and epistemology. For example, a historian clearly has skills and knowledge associated with the field, but also has a personal identity as a historian, as well as a set of values associated with the field along with a way of thinking (epistemology). Absent these last three attributes, students may know a bit of history, but have no idea why there are historians.

My first exposure to this goes back to high school when I was having trouble in history class. My teacher took me aside and told me the story of the building of the Panama Canal, the topic of his doctoral thesis. As he started talking about the intrigue behind this project, he became highly engaged, and his excitement rubbed off on me to the point that I once thought I'd follow in his footsteps! While I didn't, I remain interested in this field today, all thanks to him.

While my schools were all built around the age-old model of what my friend and school architect Prakash Nair calls "cells and bells," you can imagine how different school designs can foster the kinds of learning each of us wants for all kids.

The Danish designer, Rosan Bosch, noticed that when her own kids started school they were excited, but as the years went on, they became less and less interested in schooling—a phenomenon known to too many parents.

As a result she started designing schools, and her perspectives are expressed most eloquently in her book (Bosch, 2018), and schools designed by her firm are found in many places, including Europe, Argentina, India, Abu Dhabi, Scotland, and China.

For example, in 2011 she had the opportunity to design a school in Stockholm that children never wanted to leave (Bosch, 2013). The result, the Vittra school at Telefonplan, gives priority to the students' learning needs when planning the physical spaces of schools. By mixing art, design thinking, architecture and play, the learning environments stimulate innovative thinking and student-centered learning, supporting the pedagogical efforts and the development of 21st Century Skills. This school has 250 students.

One of her design principles is an expansion of my own work, in which she uses various learning spaces throughout the building:

MOUNTAIN TOP   CAVE   CAMPFIRE   WATERING HOLE   HANDS-ON   MOVEMENT

Bosch perspectives on learning spaces (Design by Rosan Bosch Studio)

The main distinction is the expansion of the Campfire and Life spaces and the addition of Movement. In her model, learning spaces need to enable and motivate every learner.

Everybody learns differently—and everyone needs variation. Her belief is that learning spaces need to support different ways of learning and developing skills for the 21st century. This is very important as we are already a quarter of the way through the century.

Her design of spaces is based on six principles that connect learning situations to the physical framework. Each describes a constellation for the learners' focus and interaction.

The **Mountain Top** learning situation establishes a space for individuals to address a group and let thoughts, views, and knowledge flow from one to many. The speaker or performer stands in front of an audience and becomes an educator.

The **Cave** learning environment offers a space for individual concentration, focus, and reflection. It is characterized by quietness but not necessarily isolation. Cave spaces are small, strictly defined spaces for one or two students away from areas with activities.

The **Campfire** learning situation provides a space for group-based learning situations. It trains students to work effectively in smaller teams, focus dialogue within the group, and develop collaborative skills.

The **Watering Hole** environment exploits informal spaces with many passers-through and disturbances. This is a space of disruptions where learners encounter unexpected ideas, astounding skills, and surprising knowledge that inspire and motivate them.

Hands-on is an essential design principle that adds an extra non-verbal communication dimension. It offers a link between theory and practice, mind and body, insight and play. It explains relevance, inspires and motivates learners.

The **Movement** design idea integrates movement as a natural part of all spaces. No matter a human's personality or the subject being studied, movement enhances cognitive skills and energizes the learning process.

For example, the following photo shows one of her furniture designs that is used in one of the schools she designed in Sweden.

Photo of a furniture piece in the Vittra Telefonplan school (Design by Rosan Bosch Studio)

While these learning spaces and the creativity they foster are most easily done in buildings and furniture designed for this purpose, gifted educators like you can restructure your own practice to facilitate the kinds of changes that unleash the kinds of creative learning and design you and your students explore in this book as we outline the concepts of Zero Mass Design, even if your overall school has a traditional layout.

The key for unleashing the true power of your students is to start by thinking about the kinds of activities you want them to do, and from there to think about the facilities (furniture and tools) that promote these activities. Finally, it is time to think about the building that will hold these facilities. While you may not have the freedom to start with a completely blank edifice, there is probably a lot you can do inside an existing structure by (for example) changing the furniture, adding the tools of a makerspace, etc.

This process makes you a designer as well, further increasing the utility of the ideas explored in this book.

As you progress through this book, the underlying principles of Zero Mass Design will not only empower your students, they will empower you as well.

The Basics of Zero Mass Design

The concept of Zero Mass Design can be understood by examining the relationship between two properties of a given artifact: its functionality and its complexity. "Functionality" is that property of the object which determines the number and kind of applications for which it can be used. "Complexity" includes the level of embellishment as well as the number and complexity of the individual parts from which the object is made.

One might think that complexity always increases along with functionality.

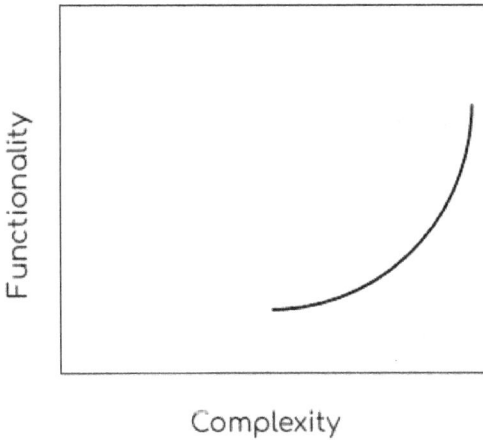

Complexity

For example, a transistor radio designed to run on battery power can have its functionality increased through the addition of an AC line power supply. This new radio can have its functionality increased still further by including a built-in tape recorder, etc. The increase of functionality through increasing complexity is not only well known, it is sometimes carried to extremes.

Occasionally one finds product offerings for which the attempt to increase functionality through complexity has reached the point of being humorous. Several examples come to mind. One such product is a fake fireplace for trailer homes which includes a built-in swing-out bar and complete stereo system hidden in the fake chimney. Another example is a stereo system which includes a built-in AM-FM receiver, tape deck, organ keyboard, and microphone.

Now, instead of pursuing the functionality curve in the direction of increased complexity, let us back up and place our attention in the realm of decreasing complexity. As complexity is decreased, functionality will decrease for a while and will then bottom out. As complexity is decreased below this point, functionality will start to increase again until, at the point of zero complexity, the functionality is increasing without bound.

Since this concept is central to this book, it must be clearly understood. The relationship between functionality and complexity is "U" shaped, and to the left of the bottom of the "U" functionality increases with decreasing complexity. Furthermore, the functionality of a design starts to rise without bound as the complexity approaches zero.

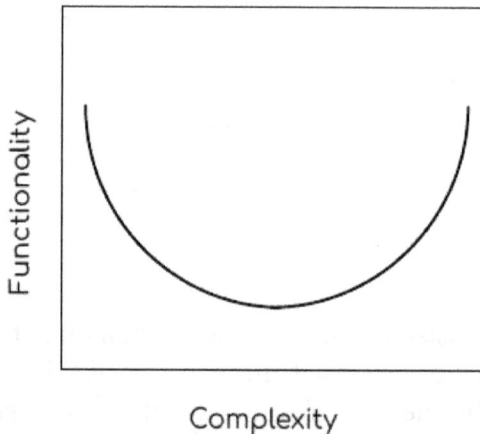

Complexity

Let's consider two examples, each of which comes from a different arm on the functionality/complexity curve. No one who has looked inside a color television set would deny that it is a complex device. From a functional point of view, a color television set is used to receive specially generated radio waves which are then decoded to produce the sound and pictures we experience while watching this device. When not being used for this purpose, the television set sits silently in the room, serving no other purpose, perhaps, than holding a vase of flowers. The color television set is a complex device with well defined (and limited) functionality. To increase its functionality, one may increase its complexity by adding, for example, a video game console, or a digital clock.

Next, contrast this design environment with that associated with the common paper clip. The paper clip is a deceptively simple device. If one believed that functionality always decreased with decreasing complexity, the number of applications for a paper clip would be far fewer in number than those for a color television, but this isn't the case. Paper clips are occasionally used to hold pieces of paper together. They are also used by children to make necklaces and long chains. Once they are bent open, their uses multiply. They become hooks for retrieving lost items which have fallen in narrow places, tools for removing ear wax, and even devices for relieving tension as they are bent back and forth during periods of stress. As an interesting challenge you might want to make a list of at least fifty applications for a metal paper clip. You'll be surprised at the result!

The point of these examples is to show that there is a fundamental difference between paper clips and color televisions, and hence between simple objects and those which are complex, insofar as each of these artifacts was the result of designing in different realms.

Since this book concerns itself with simple design, you might think that the procedures described here will only apply to paper clips and not to color televisions. That is not the case. The procedures described here will provide you with an awareness of the realm of the design space in which you are operating, and will let you operate from the simplest starting point, with complexity being added only as it is needed.

It is a principle of Zero Mass Design that one starts with the simplest design possible, even if it fails to work. The final design is then developed by evolutionary refinement from that point.

The idea of intentionally under designing an artifact may seem repugnant to you. After all, one typically wishes to design conservatively, with margin for error, and to get the job done right the first time. The problem with this conventional approach is that you end up designing in a vacuum. There is no self-generated benchmark against which you can measure your design. Furthermore, if you later find that you have overdesigned your product, you will never be able to remove from the design all of the overhead associated with the overdesigned portions.

To illustrate this, consider the story of the person who was having a house built. The contractor showed that a second fireplace could be added for an additional $2,000. The customer consented; but later, before the foundation was laid, had a change of heart. When he asked to have the second fireplace removed from the plans, the contractor told him that he would only save $1200, because of the cost associated with incorporating the second fireplace into the plans in the first place.

This experience has parallels in all fields of design.

Does this mean that cost reduction through Value Engineering is worthless? No. Reducing the cost of an existing product by finding lower cost replacements for some components and by eliminating others has utility; but it will never yield as cost-effective a product as one in which the unneeded parts never appear in the design in the first place.

To achieve minimum complexity (and maximal functionality) in a design, one should start from nothing.

Since "nothing" is a concept which is foreign to most of us, we will spend some time exploring this concept a bit. In some ways, this whole book is about nothing—"Much Ado about Nothing," so to speak. At this point it is probably obvious that, whatever Zero Mass Design may be, it is not "safe." This is true. For one thing, in order to effectively use this tool, you may have to "unlearn" a firmly entrenched design style. This is especially likely if your background is in engineering. I would like to propose that you ease into ZMD by trying it out on small portions of a larger project first. In this manner you will be able to see if it has any utility to you. If it does, you can then expand the scope of the activities to which you apply it.

As I mentioned before, if your first design works, it may be overdesigned. This is probably the hardest concept to grasp in Zero Mass Design. Most of us are embarrassed by failure—by not living up to the expectations of others or of ourselves. In order to become effective users of the tools shown here, you must overcome your fear of failure and replace it with the expectation of failure. This is not to suggest that you should become sloppy in your work for, as we will show, Zero Mass Design requires exceptional discipline on your part.

The failure we seek in our initial designs comes from starting with too simple a design. It is as though we are racing down the initial slope of the functionality/complexity curve, applying the brakes as soon as possible, and checking to see if our design has the minimum complexity required to do our job. If we are not far enough along, we can incrementally move along the curve, checking our progress along the way, and stopping when we reach our destination. In this manner we avoid the likelihood of jumping past the bottom of the curve and getting into the traditional design space in which functionality increases with complexity.

It may be useful to think of the functionality curve as being divided into two regions: the region where functionality increases with complexity, and the one where functionality increases as complexity decreases.

The region to the right side is the realm of "engineering design," and the realm to the left is that of "inventive design." These designations are in no sense supposed to suggest that engineering design is not creative, but just that engineering design procedures are more predictable and comfortable than truly inventive design procedures.

It is common, when designing something, to immediately jump to the engineering side of the curve. When ideas stream into your consciousness, they comprise a whole spectrum of incipient designs which you may not be conscious of reviewing, but which are reviewed and cataloged for utility until one for a few design(s) presents itself to you as a possible solution to the problem. The challenge in Zero Mass Design is to trap the idea generating process as early as possible—to catch the ideas before they are completely formed. In doing this you are not interfering with the creative process, only monitoring it. To accomplish this task requires great discipline since it is very easy to let the ideas run unconsciously until you are only presented with an overly complex (but perhaps quite novel) solution to your design problem. This often shows up when you bounce to the engineering side of the functionality/complexity curve.

The Five Steps of Zero Mass Design

In this chapter we will describe the five steps of the Zero Mass Design process. Each of these steps is to be taken in sequence. The steps are:

1. Problem Definition
2. Obtaining and Releasing Background Information
3. Creating a State of Relaxed Attention
4. Monitoring the Evolving Design
5. Stopping at the Earliest Possible Time

Step One: Problem Definition

It is often said that once a problem is well stated it is half solved. The creation of accurate problem statements is essential to Zero Mass Design.

The value of an accurate problem statement is shown clearly in the book Conceptual Blockbusting by James Adams (1980). In this book, he describes an interesting design problem from his own experience. As a young designer working on the plans for the Mariner IV spacecraft (the first craft to fly past Mars), he was part of a team which had been chartered to develop a shock absorber system which would work well in outer space. The function of the shock absorber was to prevent damage to the large solar cell panels as they opened up after the spacecraft was launched. The team of engineers spent large amounts of time against tight deadlines trying to construct a shock absorber that didn't have problems operating in the hostile environment of space. Oil filled dampers, for example, tended to leak in the vacuum of space and spread a film of oil over the solar cell panels, thus destroying them. Finally, as the deadline drew near with no adequate solution in sight, it was suggested that the process of opening the panels be tested in a space simulation chamber with no shock absorbers connected at all. When the test started, tension filled the room. The panels were released and flipped open by springs. They waved wildly for a minute and then settled into position without being damaged. It was thus found that the special shock absorber was not needed at all. One of the difficulties with this assignment (which resulted in the waste of a great deal of time and money) was that the problem statement, "Design a shock absorber for solar panels," carried the implicit, if untested, assumption that shock absorbers were necessary. An alternative statement of the problem, such as, "Ensure that the solar panels will not be damaged when they are opened," leaves room for more options, including the possibility that the shock absorbers may not be needed at all.

Of course, it may be argued that this latter problem statement is extremely vague, and in many respects it is. A conservative manager might take a dim view of a design team experimenting with a tremendously expensive solar panel, especially if it gets broken. Nonetheless, a broad problem statement often yields better solutions than one which is too restrictive. In industry results are often wanted quickly and the time and cost pressure often lead to a narrow problem statement that precludes a wide range of creative solutions. Part of the problem is that problem statements often come from a source other than the person(s) who are tasked with finding a solution. Part of the job of the problem solver is to push back on an overly narrow problem statement and work with those involved to reformulate it. This can be very challenging in a hierarchical corporate environment.

A principle in generating a good problem definition is that the statement should not contain the assumption of a particular solution. It takes a considerable amount of self-discipline for the person generating the statement to avoid this pitfall. Most often, the person responsible for this task will be one of the designers who will also be working on the problem. The temptation to start solving the problem before it is completely stated is quite natural, if somewhat dangerous. The creation of a problem statement is primarily a conscious analytical task. On the other hand, problem solving (in the form of design) is often a spontaneous creative task. The problem with trying to have analytical and creative processes operating at the same time is not unlike that of trying to get hot and cold water out of a single pipe at the same time. Whether you get lukewarm water or lukewarm creativity, you will not be achieving your goals.

For this reason, you must try to keep your analytical tasks as separate from your creative tasks as possible. Sometimes, defects in problem statements can be very subtle. Suppose, for example, that you are employed by a company that has started to receive price competition on their major product. A design team is often instructed at this point to "reduce the cost of the present product." This statement carries tremendous restrictions when compared with a statement which asks the team to "design a product, functionally equivalent to the one presently on the market, but which costs less." And this statement is more restrictive than one which asks the team to "design a cost effective product which will keep the company on a strong growth curve."

As a practitioner of Zero Mass Design, you will often have the challenge of taking someone else's formulation of a problem statement and recasting this statement in a form which embodies the actual goals without the inclusion of unnecessary restrictions. If one thinks of the problem statement as the foundation of Zero Mass Design, then the strength of this foundation will determine the strength of the overall solution.

Just how broad should the problem statement be? Clearly a statement can be so broad that the designer has no specific goals at all. There are many times when this may be the desirable state. If we look at what we, as individuals, are trying to accomplish with any of our present projects, each of us is probably moving along a path dictated by a "problem statement" stored away in our mind somewhere. If this statement is unnecessarily restrictive, we may pursue narrower paths than we might otherwise. Of course, if we give ourselves permission to broaden our concepts of what we might accomplish, there is the danger that we might get lost in areas of exploration which are strange to us. It is my belief that, even if we ultimately decide to work within a narrowly defined range of our capabilities, it is valuable to go through a planning exercise which brings these other possibilities to our consciousness.

It is one thing to make a conscious decision to not follow a certain path, and altogether another thing to miss a path because we didn't know it was there.

As a designer, you have the challenge of taking a problem statement and ensuring that it is as broad as possible, while still stating the objectives clearly. Let us follow a problem statement through this broadening process to the extreme.

Problem statement as given:
1. Trim $50 off the cost of our model 401 computer printer by replacing the metal parts with plastic parts.

Derived statements:
2. Reduce the cost of our model 401 computer printer by $50.

3. Design a computer printer which is functionally equivalent to the 401, but which costs at least $50 less.

4. Design a product which will compensate for the price competition on the 401 printer and which will keep the company on a strong growth curve.

5. Design the least expensive printer you can.

6. Design a printer that is cost competitive and functionally better than the 401.

This opens up the possibility of improved and differentiated designs. Perhaps in analyzing the competitive problem, the best market solution might be a superior printer!

Statements 1 through 6 cover an exceptionally wide spectrum of opportunities. As a practitioner of Zero Mass Design, your goal will be to move towards the broadest problem statement possible.

The representative of the company who presented you with statement number 1 has the goal of staying as specific as possible. This is probably because he or she feels that this will help focus the design task and thus get you to a solution faster. As a designer, you have the challenge of gaining as much latitude as possible in which to design. Self-employed designers, for example, usually spend at least a fraction of their time on the solution of problem statement 5. This does not result from an act of defiance, but is the natural consequence of spontaneous invention. A designer can no more be asked to invent only in a rigidly defined area than he or she can be asked to breathe only on even numbered days of the month. Since spontaneous inventive acts occur anyway, there is an increased likelihood that a broad problem statement will find a solution faster than a narrow one.

One of my colleagues at Xerox PARC, Roy Lahr, was a master of this task. His primary job was advising the Xerox mergers and acquisition team on what companies Xerox might acquire. As a gifted inventor in his own right, Roy had a keen sense of which companies had product lines based on the broadest problem statements possible. In some cases, I was put on loan to some of these companies to help them redefine parts of their products along the principles described in this book.

One of the things you might notice during the process of rewriting problem statements is that, as the statements become broader, they gain some similarity with problem statements from other projects. At some point (statement 5 above) problem statements converge.

Just as your spontaneous inventions may apply to a broad problem statement, it is also likely that these inventions may apply to several problems at the same time. This phenomenon makes it desirable to be working on several problems at once, since ideas revealed during the solution of one problem may have applicability to another problem area. This synergistic interplay of problem areas is enhanced through the conscious effort of broadening the problem statements to the extreme.

Once you have created a list of several problem statements in which each successive entry is broader in scope than the one which preceded it, you should try to gain permission to work on the problem described by the broadest acceptable description. Just as your first creations should be underdesigned, you might want to select a problem statement which is just a little too broad to be acceptable to the person who generated the original statement. The reason for this is that the ensuing dialog between the two of you will result in a clearer definition of which portions of the project are flexible and which portions are not. It is important to be on good terms with the individual with whom you are negotiating. If this person genuinely feels that you have the project's success as a personal goal, you will get much farther than you would if you are perceived as someone engaged in back-biting criticism. Your task is to gain as much space in which to do your creating as you possibly can. This is a non-trivial responsibility and one which has to be carried out tactfully. There was a time, for example, when I was consulting on a large disk drive on which the cable was vibrating so much, the thin conductors broke. The client wanted me to redesign the cable, but I was able to convince him to try gluing the cable in place with a drop of super glue instead. This worked perfectly and solved the problem quickly for negligible cost.

Step Two: Obtaining and Releasing Background Information

After settling on an acceptable problem statement, you are ready to pursue the next step of Zero Mass Design. This step is the one in which you absorb all the background information you can. Much of this information can be sought in specific places. For example, if you are designing a new printing press, you might want to read online journals on Mechanical Engineering. Of equal (and usually greater) importance are the chance discoveries you make while pursuing other information resources. As a child I read anything I could see. Many breakfasts were spent reading the backs of cereal boxes, just because they were on the table. Today I subscribe to about twenty online resources a month. Some of these are general interest and news periodicals, and others are science and technical journals which deal with advances in fields of interest to me (computers, space exploration, etc.).

Years ago my favorite magazines were the trade journals (covering everything from retailing to circuit design) which were sent free to qualifying subscribers. These magazines were paid for by their advertisers. As a result, the overwhelming majority of each magazine was devoted to lavish full color advertisements for such items as capacitors or power supplies or ball bearings. Rather than view advertisements with disdain, they could be a tremendous source of ideas—probably because of their carefully designed layouts which were geared for maximum visual stimulation. Of course, now that print has given way to online resources, the advertisements you see are more focused.

Obviously you can't expect to read every resource completely. One approach is to spend about five minutes with each one, glancing briefly at each page on your screen. If an article (or advertisement) catches your eye, and needs to be studied in detail, bookmark it to be read later. During this scanning time it is important to not be thinking specifically about anything—and especially not to be thinking about the specific problem you are working on. If something of value appears, the magic AHA! will make sure it's remembered.

More traditional reading styles should be reserved for literature searches on the history of the problem you are solving. If you are working on anything of significance, it is quite likely that you will uncover hundreds of previous attempts to solve the problem. Whether your search leads you to the Patent Gazette, or to other such online journals, you will find that lots of ideas have already been tried. It is important not to be discouraged by articles which stress the difficulty of your task. It is most likely that you will find a totally new approach to the problem—an approach which circumvents the problems found by others.

As with the online journals, you should not pay too much attention to the immediate applicability of any idea you come across. You are not looking for hints or specific help of any kind—you are just getting background information. While reading articles and scanning through articles can provide you with a lot of background material, there are many other valuable ways to get information. Among my favorites are large toy stores, hardware stores, lumber yards, discount stores and catalog showrooms. I have spent many hours rummaging around plumbing departments only to find that the existence of a seven cent plastic pipe fitting ended up saving many hundreds of dollars in the design of a custom part for some project I worked on later.

Of course, today there are lots of great informational resources and catalogs online. While I preferred the paper documents better, I've learned to peruse the online resources effectively, especially if I know the kinds of things I'm looking for.

Depending on your needs and personal style, you might want to rummage around scrap yards and garbage dumps. The goal here is not to be a scavenger of materials, but to be a scavenger of ideas. What ideas didn't work? What products broke down? Why? Keep in mind that if you were picking through garbage dumps that were many thousands of years old, you would be engaged in the respectable profession of archeology. By searching through more modern preserves you are just accelerating the process a bit.

While no less supplied with "garbage," I find industrial trade shows to be a tremendous source of information. If you live in or near a major metropolitan area, you probably have easy access to a few such shows every year. Many of these have little or no admission fee, but are limited to those people working in the technical area covered by the show. If you really want to attend one of these shows, you will find entry very easy.

If money is no object, one of the most spectacular industrial trade shows in the world is the Hanover Fair, held every Spring in Hanover, Germany. Housed on a fairgrounds the size of a small city, this ten day fair has millions of square feet of exhibits historically covering everything from desk calendars to railroad trains. Companies from countries all over the world exhibit their products for the 250,000 or so attendees. For pure stimulation of the creative process there is nothing like it. Over time the focus of this fair has evolved, and in 2022 the main areas showcased include AI and machine learning, decarbonization, the circular economy addressing global climate change, hydrogen and fuel cells, among others.

If your focus is on technology and your travel budget is restricted, the Consumer Electronics Show held in Las Vegas every January is a perfect place to gather ideas.

Information is wherever you look for it. An article here, an exhibit there, a toy or chance comment elsewhere—all these can combine synergistically to synthesize a new idea which becomes the solution to the problem you are working on.

Once you have exposed yourself to these various resources and are ready to solve your problem, you may wonder what you are to do with all the background information you have gathered. In Zero Mass Design what you do next is very simple.

Forget it.

Step Three: Creating a State of Relaxed Attention

Once I was brought into a "fire-fighting" situation involving the manufacture of a new type of computer keyboard. The research prototype worked perfectly, but the process didn't make the successful transfer to manufacturing. Faced with pitifully low yields (and a major slip in production of the computer which used this keyboard), I was told to drop all other projects and focus on this one until it was fixed. I spent one day at the site getting background information, and spent another day reviewing my information on the manufacturing technologies associated with this type of product. On the third day, faced with nervous product managers who kept an hourly watch on me—looking for any sign of an early solution—I did something quite strange.

I went to the beach.

At that time I lived about an hour's drive away from a secluded beach overlooking the Pacific Ocean. Perched up on some rocks, away from all other people, I was able to just sit and stare at the water without being disturbed by people or telephones. The only sounds I heard were the cry of the birds and the roar of the ocean as it beat against the rocks.

The next morning, I realized that I had found the solution to the manufacturing problem. This was confirmed in the laboratory in a few days. The solution was quite simple (too simple, in some people's eyes), and rigorous life testing of the new keyboards was carried out. The results were positive. Not only did the new process solve the manufacturing problems, but it eliminated several labor intensive steps and resulted in a significant cost reduction as well—even though reduced cost was not one of the project objectives.

The time I spent at the beach, clearing my head of extraneous thoughts, gave my creative process the freedom it needed to generate a solution for me. At no time did I sit down and say "OK, let's make a list of alternative solutions." I didn't even "think" about the problem at all. The solution just presented itself to me at breakfast.

This happens so frequently that a relaxation period is an essential part of Zero Mass Design. Each of us probably lives within an hour's drive of at least one secluded spot where we can be free from distractions. If you like forests, then by all means go to one. If you like parks, zoos, or beaches, then go where you want. The important function served by this place is to allow you complete isolation from day-to-day problems so that you can unfocus your mind. You should try not to think about anything. Avoid all distractions. This means avoiding all other people, radios, pinball machines, cars, food, or anything that causes you to "think." You should stay at this place until you feel relaxed, but not stay so long that you become excessively tired or famished. The cleansing feeling you seek has been called "relaxed attention." You are in a relaxed state, but you are still aware of the rest of the world.

The concept of relaxed attention is an important part of the creative process as pioneered by Professor Robert McKim in his book, *Experiences in Visual Thinking* (McKim, 1980, 38-44). This relaxed state moves us toward "effort"—the direction of physical and mental energy toward the tasks of effective visual thinking. To distinguish effort from trying, we define trying as the inappropriate application of energy to tasks. The artist should not prematurely determine the direction which their artwork will develop. Visual thinking should NOT be rigidly controlled by verbal processes in the mind. The goal is to move us toward a state of "flow," the ecstatic feeling that everything is going just right when the individual is totally immersed in a creative act. The flow experience is also characterized by a sense of losing contact with time and the external world. It is facilitated by a physiological state of relaxed attention. This concept was originally explored by Professor Mihaly Csikszentmihalyi (2008).

You may feel that you just can't take a day off to spend at the beach in the middle of a major project. After all, suppose nothing comes of this escapade. All you have accomplished then is to lose a day on the project and incur the wrath of your colleagues whose temperament is determined by the extent to which their necks are on the line. Well, as stated before, Zero Mass Design is risky. Sometimes the process works and sometimes it doesn't. My experience is that the successes are so striking that the occasional failures are tolerable.

If you really feel that you can't leave your office, take some relevant books and papers to the library, or some other acceptable quiet place, and try achieving a state of relaxed attention there. With a little practice you will be able to stare at almost any book without thinking about what you are seeing. Of course, only you will know that you aren't busily studying up on the problem, and your deception should be quite successful.

Another useful technique is mindfulness meditation (Sokolov, 2018). This is very easy to do and can be done anywhere (your office, a quiet place in the library, even during a commute on public transportation). This technique is widely practiced and with time helps free up the mind to gain a state of relaxed attention.

Step Four: Monitoring the Evolving Design

The evolution of a new design is something which you should monitor quite carefully. If your previous experience as a designer has been to work with very complex and elaborate ideas, you should make a special attempt to capture new ideas as soon as they emerge. Capturing an idea is not as easy as it sounds. You may have had the experience of waking up with a vivid image of a dream still in your head, and finding that all conscious memory of the dream evaporated as soon as you started to describe it to someone else. If you have an idea which is sufficiently well formulated that you are able to discuss it quite freely, it may have already undergone a significant amount of unconscious embellishment, and thus end up being much more complex than it needs to be.

The goal is to find a way to capture your ideas when they are still in the "dream-like" state, and before they have had a chance to be made more complex. You must try to catch the ideas when they are too simple to work, and then carefully carry them to the edge of success. While each of us has to find our own idea-capturing mechanisms, some designers find that rough sketches and crude models are a better capturing medium than the written word. There is something about writing which tends to drive fragile wispy thoughts away. While one probably shouldn't travel without pen and paper handy, sometimes ideas can be captured in wordless form by going into the shop and shaping an object out of wood or styrene foam. Once the idea has been captured this way, you are then free to "think" about what you have just built, and to examine what relevance, if any, it has to the problems on which you are working at the time. If you start thinking about the design before you capture it you may drive it away completely. The model or sketch (which is often meaningless to anyone else) becomes the trigger to refresh the original thought in your head long after it has made its first appearance.

While you might often "see" or "feel" your inventions, you might also "hear" or even "smell" them instead. A small tape recorder (which you can always carry with you) is a most valuable tool for catching your repetition of fragments of the inventions you hear. Speaking is so much faster than writing, that you will often be able to get enough words captured to allow reconstruction of the idea at a later time. Pay no attention to forming complete sentences; as with drawings, your words may lack sufficient coherence for others to make sense from them, but you will be able to reconstruct the idea from these fragments yourself. Later on, I'll share one of my inventions that was first written on the bottom of a Pampers box during a diaper change for my son during the middle of the night.

Whatever the means, you should try to capture enough of your thoughts in a form that will have enough of the essence of your ideas to allow them to be intelligible to others.

Once the idea has been captured, you should think about its applicability to your problem. Build a working model if possible. If this model doesn't suit your needs, try to guess how close you are to a solution. If a simple modification is all that is needed, you are in good shape. If the idea doesn't seem to work at all, then you might have to start over again.

Above all, don't hang on to an idea once you have shown it to be worthless. I once invented a digital recording process which (without my knowledge) violated the Second Law of Thermodynamics (such processes can be used to build perpetual motion machines of the second kind). Needless to say, I was most impressed with my ingenuity until its flaw was pointed out to me by a kind friend. I felt devastated! It took me about a week before I was able to "let go" of the bad idea and start over. While it is very common to become attached to your inventions, you need to know how to step back from them and apply critical appraisal of their worth. This is a hard skill to master, but one which is essential to your success as a designer.

As R. Buckminster Fuller said in his book, *Ideas and Integrities* (1969, p. 31):

> *We must learn to do all the little wrong things first in order to learn by direct experience that we must take broad, comprehensive and incisive responsibility in the formulation of our overall strategies.*

Step Five: Stopping at the Earliest Possible Time

Once you have homed in on a solution to your design problem you need to know when to stop the design process. There are two arguments in favor of stopping this process as soon as possible. The first of these is to ensure that your design has the minimum complexity. Look very closely at your prototype. Have you used a nut, bolt and two washers when a piece of double-sided foam tape would have worked as well? There are lots of ways that complexity can creep into a design, and they should be avoided at all costs. Remember that it is easier to add needed complexity than it is to remove excess complexity.

The second reason for stopping a design as soon as possible is applicable to those whose designs are to be implemented by others. I have never met a design team yet which was willing to accept someone else's ideas without modification. Personal pride doesn't allow most engineers to take an idea from a prototype to a product without "improving" it. While some improvements may be warranted, it is not uncommon to see a perfectly simple and beautiful idea "improve" beyond recognition because of someone's desire to receive credit for their ingenuity as well. I have found that the best way to handle this problem is to bring the other members of a product team into the project as soon as the rough ideas are ready to try out. Quite often these people's contributions can be quite helpful in the evolution of the design. Once their ideas have been incorporated, they will be most reluctant to make extra changes later on.

Like many other aspects of Zero Mass Design, the fifth rule is quite simple: When you are done, stop!

Fuller's Thoughts on a Generalized Design Process

While R. Buckminster Fuller's work on the process of design differs from ours in some ways, his ideas overlap a bit with ours, so this chapter gives a brief view of his thinking as summarized by Michael Ben-Eli (Design Science: A Framework for Change, 1970).

A complete design process comprises a number of distinct phases, or steps, each incorporating a cluster of a different type of activities. Five essential steps appear to be consistently involved along the Universal Design Spiral.

They include the following:

Intention

    Formulation

        Realization

            Operation

                Transformation

Intention
This is the initiating step. It may emerge with a vague recognition. It represents an initial impulse, a hunch, an inspiration — before analysis or a fully developed rational argument. It is largely intuitive and is essential to the initial energizing of motivation. It springs from a cumulative synthesis of all previous experiences.

Formulation
This phase involves the increasingly sharper formulation of a concept. The focus is on defining a purpose, a galvanizing vision, a mission, a strategy, an approach. Early rounds of research and analysis are

involved in articulating and shaping all of these. A plan, a program, a detailed blueprint, is the end result.

Realization
In this phase, the conceptual blue-print drives a reiterative process of modeling, prototyping, testing, refinement and debugging. It culminates with the production, construction or other forms of implementing a new design.

Operation
This phase moves from implementing a concept to all the aspects associated with routine use. Issues of maintenance, service, recycling and the like. Where relevant, it also involves incorporation of adaptive refinements following actual experience with use.

Transformation
This step represents a discontinuity relative to a given design cycle. A new intention emerges in response to fundamental change in the context and a new cycle begins. Unlike changes which characterize adaptive refinements, where despite a series of improvements and adjustments the general framework remains unchanged, this step represents a major shift, a clear break from the past. For example, a shift from a horse drawn carriage to a motor car, from wire to wireless technology, from surface to air transportation, from an agrarian to an industrial society, from classical to modern painting, or from centrally planned to free market economy.

Additional Aspects of the Design Process
A number of additional characteristics define the design process:

- The process is adaptive, meaning that elements in all of the five essential steps continuously co-define and inter accommodate, in relation to each other and with respect to the context as a whole
- The process is reiterative, requiring continuous adjustments and re-adjustments as advances are made at each step

- The process is recursive, meaning that a similar circular structure that characterizes the whole reappears within each one of the individual steps

A few additional comments about aspects of the design process are instructive:

- The recursive aspect, often a source of confusion, is of important logical and practical significance. It means that "purpose," for example, requires formulation and reformulation with each step, where it is articulated in a manner appropriate for that particular step. It would thus be defined at different levels: as a general intent; as a particular objective; as the guiding essence of a strategy; and, as the series of specific goals necessary for implementation.

- The design process as a whole is often represented as a sequence of clearly differentiated logical phases. In reality, it operates more like a non-sequential network of dynamic, self-organizing, multiple, interacting events.

- The process can be instantaneous or it may need to be organized over a considerable period of time.

- The process may involve a single individual or it may require the integration of multiple teams representing, at different stages, different capacities and expertise. The latter raises the issue of effective management of the design process as a whole.

- Finally, a point that is all too often neglected particularly in broader socio-economic aspects of human affairs, namely, that the design process, for each given case, needs to be the subject of deliberate design in itself.

Fuller wrote extensively on the design process and the effects of bad decisions on society (Fuller, 2008)

While Fuller developed his own methodologies, many of his creations embody the underlying essence of Zero Mass Design. He felt that failure was important because if your design works the first time, you just haven't tried hard enough to establish a minimal design. If your design works, back up and simplify it until it doesn't work. You might find in the failure a different path to take that you haven't considered.

Examples of Zero Mass Design

There are myriad examples of Zero Mass Design; we need only to look for them. Once they are found, we can see what it is that makes these designs so powerful in their simplicity. This chapter is a collection of a very small sample of such designs. To my knowledge, only some of the designs described here were actually arrived at by the process described earlier in this book. There are many paths to a given result, and it is with the results and their implications that we will be concerned in this chapter. By analyzing these examples you may gain some deeper insights into the essence of extreme simplicity in design and thus be better able to identify this essence in your own work. Each of the examples is self-contained so that you can read them in any order. You are encouraged to read all of the examples, even if only one or two of them are remotely associated with your fields of interest.

The Four Color Map Theorem

Tensegrity Structures

Upcycling

The $750 Poem

License Plate Linguistics

Twenty Questions

The Tapered Resistor

The Resistive Touch Tablet

A Computer for Everyone

Spring Reverb

Traditional Japanese Housing

3D Printing

IKEA Furniture

The Four Color Map Theorem

In mathematics, the four color problem, or the four color map theorem, states that no more than four colors are required to color the regions of any planar map so that no two adjacent regions have the same color. Adjacent means that two regions share a common boundary curve segment, not merely a corner where three or more regions meet. It was the first major theorem to be proved using a computer and was proved in 1976 by Kenneth Appel and Wolfgang Haken. The problem itself was known for hundreds of years, and a formal proof eluded many mathematicians world-wide.

In his groundbreaking book, *Synergetics*, R. Buckminster Fuller stated that this pivotal theorem could be solved by inspection:

> *Polygonally all spherical surface systems are maximally reducible to omnitriangulation, there being no polygon of lesser edges. And each of the surface triangles of spheres is the outer surface of a tetrahedron where the other three faces are always congruent with the interior faces of the three adjacent tetrahedra.*

> *Ergo, you have a four-face system in which it is clear that any four colors could take care of all possible adjacent conditions in such a manner as never to have the same colors occurring between two surface triangles, because each of the three inner surfaces of any tetrahedron integral four-color differentiation must be congruent with the same-colored interior faces of the three and only adjacent tetrahedra; ergo, the fourth color of each surface adjacent triangle must always be the one and only remaining different color of the four-color set systems.* (Fuller & Applewhite, 1982, section 541.21)

As stated, this proof applies to planes (the normal configuration of the kinds of maps we are likely to encounter). In this case the simplest shape that encloses an area is a triangle, and a triangle has three edges. If the triangle has one color, then each of the three shapes bounding the triangle has its own color, giving four colors in all. This "proof" applies to any planar surface, even if the boundaries between regions are infinite fractals.

While it is true that one could envision complex maps for which the solution is not true (toroidal maps, for example), Fuller's proof is based on the kinds of maps we use in our world, and for these maps, his ideas work perfectly.

When thought about in the context of Zero Mass Design, Fuller's work clearly applies the core ideas, especially if we refine the question to its basic form. The four-color map theorem proof has baffled mathematicians for ages. By taking a simple approach of realizing that a triangle is the simplest shape with an inside and an outside, Fuller argued that any two-dimensional map could be represented by a web of triangles each of which borders another one on three sides (giving 3 colors maximum) plus the interior of the triangle, yielding the fourth triangle. This simple formulation follows the principles of Zero Mass Design.

Tensegrity Structures

For thousands of years, large structures were made by stacking structural blocks on each other. From the great pyramids, to Gothic cathedrals, and, likely, to your own home, the dominant construction method was based on elements in compression.

This form of construction is so common that building blocks have been a popular toy for hundreds of years.

The stability of large compressive structures has been advanced by the invention of structural elements like the flying buttresses used in Gothic cathedrals.

Flying Buttresses in the Cathedral of Notre Dame, Paris. Photo by Jean Lemoine from Béthisy-Saint-Martin, France - Flickr.com, CC BY-SA 2.0.

These innovative structures allowed for thinner walls and the increased use of stained-glass windows. They also add tremendous complexity to large buildings.

In the early 20th century Buckminster Fuller's observed that some materials (e.g., cement, stone) are strong when compressed and weak when pulled, and others (e.g., wire) are strong when pulled, and weak when compressed. A tensegrity system is established when a set of discontinuous compressive components interacts with a set of continuous tensile components to define a stable volume in space. Through the design of tensegrity polyhedra, he was able to craft structures where every element was either in pure tension or pure compression. The result is a stable structure that can be replicated to make very large items like radio antenna towers, for example. Because the compressive elements are not stacked on one another (in fact, they don't even touch, there is no way for them to be stressed to their breaking point.

For example, the tensegrity octahedron shown below was built using 3D printed plastic compressive struts and rubber band tension pieces:

One of the interesting qualities of tensegrity structures is that when they are stressed (by wind, for example), the compressive elements are pressed more, and the tensile parts are pulled more (Saidani & Remise, 2002). Even though the structure may look a bit flimsy at first glance, it is actually quite strong. Tensegrity modules can be stacked vertically, or cantilevered, and still work well. For example, the following sculpture is 30 meters tall.

Needle Tower II (1968) by Kenneth Snelson as seen from the bottom (CC BY-SA 2.5 nl)

Upcycling

One million plastic bottles are discarded every minute. As landfills reach their capacity, plastic bottles are finding their way to the oceans, leading some researchers to predict that, by 2050, the oceans will have more plastic bottles than fish. (Data on this topic can be found at earthday.org.) While various organizations are working to reduce the amount of single-use plastic around the world, this section will talk about some Zero Mass Design ideas that can repurpose waste in productive ways—the process of "upcycling."

Upcycling differs from recycling in that the waste materials can be turned into useful items by combining them {in their original state or parts of them with other things individuals can make on (for example) a 3D printer.

I first saw this in action in Brazil a few years ago when I spotted a Christmas tree made from empty 2-liter soft-drink bottles.

Upcycled Christmas tree photographed in Olinda, Brasil

Ideas for upcycling projects can be found in many places online including the "my mini factory" website (myminifactory.com).

For example, this art deco vase by Conor Devine uses an old lightbulb and a simple 3D printed bracket.

Photo from myminifactory website

This site has many hundreds of 3d printer designs for upcycling waste products, and even has a category for these designs.

Categories

ACCESSIBILITY	ARCHITECTURE	BOARD GAMES
BUILD A 3D PRINTER	EDUCATION	FAN ART
FASHION & ACCESSORIES	GADGETS & ELECTRONICS	HOME & GARDEN
JEWELLERY	PROPS & COSPLAY	RC CARS
SCAN THE WORLD	SPARE PARTS	SPORTS & OUTDOOR
TABLETOP	TOYS & GAMES	UPCYCLING

Anyone who wants can add their designs to this category.

While upcycling won't significantly reduce the amount of waste in the world, it may well result in ways to help you design items that meet practical goals and meet the design criteria of Zero Mass Design. This would make a great design project for anyone teaching ZMD in a school.

The $750 Poem

In the autumn of 1965, Aram Saroyan wrote a poem, shown in its entirety below:

lighght

This seven letter poem became the center of a controversy which has contributed to the attempted dismantling of the National Endowment for the Arts. Editor Robert Duncan selected this poem (which he felt to be most striking in its simplicity) for the American Literary Anthology. A grant of $750 was made for each poem which appeared, regardless of its length.

No sooner did this poem appear in the anthology that a clamor arose among conservatives in the House and Senate who decried the "waste" of government money on such a "simple" effort. In 1970, Representative Wiliam Sherle denounced this poem from the floor of the House as a misuse of public money at the rate of $107 per letter.

When running for the Presidency, President Regan used this poem as an example of boondoggles in the arts program, and then used it as justification for slashing millions of dollars from the budget for the National Endowment for the Arts.

There are many complex issues surrounding government support of the arts, and this is not the place to discuss them. The main reason for elaborating on the arguments surrounding Saroyan's poem is that the criticism leveled against the work was based not on its quality, but on its simplicity. It is as if the detractors were saying that nothing that simple could be worth so much. This criticism has been leveled against other examples of Zero Mass Design as well.

Is Saroyan's poem too simple to be good poetry? Each of us must find our own meaning in any poem. Poetry doesn't have meaning all by itself; we bring meaning to poetry.

Clearly the poem had meaning to Robert Duncan, or he wouldn't have selected it for his anthology. Clearly it had meaning to Saroyan, or he would never have published it in the first place.

Given the response this poem has elicited from many government officials, it must even have some meaning for them.

License Plate Linguistics

States which allow personalized license plates for automobiles have accidentally created a golden opportunity for Zero Mass Design. The freedom to create within the constraints imposed by the plate's ability to hold only a few letters and numbers has proven to be a most enjoyable challenge for some people. While license plates such as MY CAR, HERS, etc. are not terribly inspiring, those which get the most attention are the ones which incorporate multi-word sayings in the space of a few characters:

10SNE1

An old joke asks what the woman said when she looked in the pail:

OICURMT!

The idea of minimal representation can be seen in the poem:

Little Mary put on her skates,

Upon the ice to frisk.

Her friends thought she was most brave

Her little *

Puns (such as the use of the asterisk above) are quite common among inventors and other creative people. The twisting of words and phrases to create new meanings for old sounds has much in common with the twisting of ideas to create new perspectives which lead to invention. Punsters are probably using the same creative energies as other designers. This does not mean, however, that you must subject your friends to horrible puns in order to show that you are a good designer. Texting abbreviations are another more recent form of this kind of creativity (CUL8TR, BTW, IDK, LOL, ROTFL for example.)

While the sparse representation of phrases by the use of letter sounds has been the subject of humor for ages, a more visual form of this type of humor was published as a regular feature in the magazine *Games* as "Wacky Wordies."

Today, a large collection of these are published on the Web by David Pleacher (bit.ly/3E3RWq0) such as:

GROSSLY
STATED

or

0
BS
MS
PhD

Clearly there is no end to the possibilities for this type of game. As an exercise you might want to design a few of these yourself. As a final challenge to this set of examples, you might want to generate a profound saying using the fewest number of characters. Since this is a book on philosophy, I will submit a one letter saying of importance to most philosophers:

Y

Answers	GROSSLY STATED
10SNE1	*Grossly overstated*
Tennis anyone?	
	0
	BS
OICURMT!	MS
Oh, I see you are empty!	PhD
	Three degress below zero

Twenty Questions

Back in the early 1980's I was asked to speak at a computer conference on the topic of Artificial Intelligence. In 1950, the scientist Alan Turing proposed a definition of computer intelligence based on whether or not the user was unable to tell if she was conversing with another human or a computer. (Turing, 1950, 433)

I said that we'd demonstrate a game, Twenty Questions, and then show how my program worked. I asked for a volunteer who was good at this game, and typed his responses into my computer. The resulting dialog looked something like this:

```
Person: Is it larger than a breadbox?
Computer: No.
P: Is it alive?
C: Yes.
P: Is it a person?
C: No.
P: Is it a vegetable?
C: Yes.
P: Is it a legume?
C: Yes
P: Is it a kidney bean?
C: No.
P: Is it a chickpea?
C: Yes.
```

Success after seven questions! Of course there were some pauses during the game play, presumably to parse the person's question.

So how did the computer play the game?

In fact, the computer had no idea what the answer would be. All the program did was check to see that each query ended with a question mark. Then it looked at the last letter in the query. If it was a vowel, it typed Yes, and otherwise typed No. The trick was choosing a player who was good at the game. This person was, in fact, playing against himself.

When I showed the program listing, some in the audience claimed that I wasn't showing the actual program. They thought the actual program would have been many pages in length. This illusion was aided by waiting a random period of time after each question as if the computer was furiously trying to parse the query. The whole program, originally written in LISP and recently rewritten in the Logo version called Lynx (lynxcoding.club), is shown below:

```
to 20_questions
make 'counter' 1
repeat 20 [
question sentence 'Enter question' :counter
make 'my_var' answer
show :my_var
ifelse (last :my_var) = '?'
[ifelse member? (last butlast :my_var) 'aeiou' [wait
random 5 show 'Yes'] [wait random 5 show 'No']]
[show 'Please ask a question.']
make 'counter' :counter + 1]
end
```

The title of my presentation was "What is the Difference Between Artificial Intelligence and Natural Stupidity?"

Needless to say, I was never invited back to that conference.

I'm including this example to show that the principles of ZMD sometimes can even apply to the writing of software.

The Tapered Resistor

When my son, Harvey, was born, I was in charge of late night diaper changes. Bleary-eyed, I'd respond to his cries and get his bottom cleaned, dried, and freshly wrapped in a clean diaper.

One night, while taking care of him, an idea popped into my mind that I sketched out on the bottom of a Pampers box.

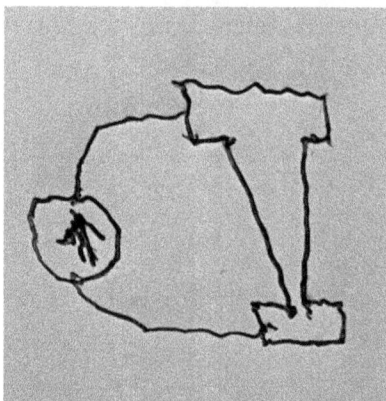

The next morning, I looked at this sketch to try to figure out what it was. It had a tapered element and appeared to be connected to a current source. My first thought was that I'd recently purchased a gas flow meter that ran gas through a tapered tube containing a small ball. The ball moved up the tube by a certain amount based on the flow of gas in liters/sec.

But the presence of an electrical current suggested something else. If the tapered region was a resistive film, then the passage of current would heat up the narrower region more than the wider region. If the resistive area was coated with a material that changed color on reaching a certain temperature (a thermochromic liquid crystal film, for example), then this device would show a colored line whose length was related to the current flowing through the resistor.

Within a few days I built a prototype using silk screened graphite-based ink to make the tapered resistor on which a thermochromic film was glued. I demonstrated this invention to my colleagues at Xerox, and a patent application was made. In October, 1978, patent 4,121,153 was granted for my Tapered Resistor Meter (figure below).

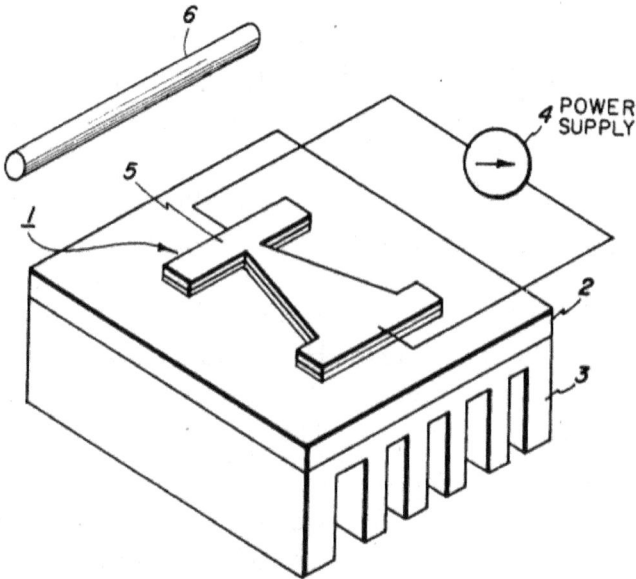

Prior to this invention, electrical current meters were expensive, fragile and bulky. While the tapered resistor meter was cheap and robust, it was not as accurate as traditional meters, but there are many applications where pinpoint accuracy is not required. The fact that this device could be made for pennies instead of dollars, meant that this device was disposable.

In fact, in the late 1990s the tapered resistor was built into Duracell alkaline batteries to allow customers to insure that a battery had enough power to be used properly.

The tapered resistor technology used in the Duracell was licensed from Kodak's patent, 5,059,895, issued in November, 1991. This invention is identical to the one I invented thirteen years before. I called the good people at Xerox and suggested they sue on our behalf. The response was that lawsuits were expensive and that damages would be hard to assess since the meters were being given away by the battery companies for free.

In the context of this book, this invention was the result of several aspects of the Zero Mass Design process and this makes it a fine example for inclusion in this book, independently of its limited commercialization. The idea came to me in the middle of the night when your brain was not in rational problem solving mode. In fact, the following day it took me a while to realize what I'd invented. I was able to realize that the simplicity of the idea was its strength.

I was far from the first to have a breakthrough outside a rational problem solving mode. Elias Howe, was an American inventor who wanted to refine the concept and design of sewing machines of the 1800s. He struggled with the key component of an automated stitch until it came to him in a dream. Howe's original idea was to follow the model of the ordinary needle, and have the eye of the needle located at the heel. This did not work for what he wanted to accomplish.

In his dream, he was being chased and attacked by cannibals with spears that had holes in their tips. It was after this dream that he realized that the needle in his invention must have a hole in the tip to put the thread through it, which led to an innovative, efficient, and functional design used in modern day sewing machines.

The Resistive Touch Tablet

The origin of this invention starts with a visit of Alan Kay to MIT in the late 1960's where he met with. Professor Seymour Papert, one of the inventors of the Logo programming language designed to help all kids learn mathematical thinking by building their own programs.

Alan, soon to join the Xerox Palo Alto Research Center (PARC), thought that, if kids were going to create their own programs, they needed their own computers on which they could play around with powerful ideas.

Describing the idea as "A Personal Computer For Children of All Ages," Kay created the Dynabook concept to embody these proposals. An early history of this concept was published as an article by Alan Kay and Adele Goldberg (1977).

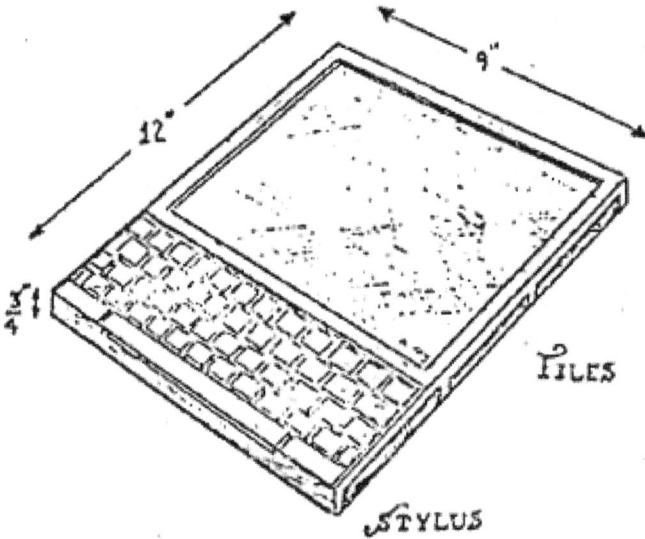

Sketch of original Dynabook concept from Kay and Goldberg paper

Kay's concept was created two years before the founding of Xerox PARC. The ideas contributed to the development of the Xerox Alto prototype, which he originally called "the interim Dynabook." It embodied all the elements of a graphical user interface, or GUI, as early as 1972. The software component of this research was Smalltalk, which went on to have a life of its own independent of the Dynabook concept. The technologies like thin flat screens had yet to be invented at that time, so the Alto was, in fact, a desk-sized computer with a CRT display and a mouse pointing device and separate keyboard—in short, a far cry from the tablet concept envisioned by Kay, and ultimately made real as the iPad in 2010 along with several Android-based tablets around the same time (Kay, 1986).

One of the desired features of the Dynabook was a transparent touch screen on which users could draw and point to objects, including virtual buttons.

In the early 1970's, several types of digitizing tablets were on the market. They used a variety of position-sensing technologies and were very expensive, ranging up to $3,000 for high-resolution models. In addition to prohibitive cost, these tablets were not transparent, thus making them inappropriate for the Dynabook.

Rather than think about modifying existing designs, I chose to start from scratch. I realized that if I had two resistive sheets of flexible plastic coated with indium tin oxide (a transparent resistive material) and had a good conductor running horizontally at the top and bottom of one sheet, and another similar sheet with the conductive strips running vertically on the left and right edges of this sheet, then if the resistive faces were spaced closely to reach other, the sheets would be connected at a point by touching the top sheet with a finger or stylus and pressing the two sheets together at this point. Electrically this would represent two potentiometers whose wipers were connected to each other when the top surface was pressed against the bottom one. This was in clear contrast to the tremendously expensive tablet digitizers that used elaborate technologies to provide very highly accurate readings of coordinates.

The decision to use these plastic sheets with transparent resistive coatings came from the use of these sheets in heated defrosters in fighter aircraft windshields. The fact that these sheets were inexpensive to make fit into the ideas of Zero Mass Design quite nicely.

The main point is that I started from scratch with my design.

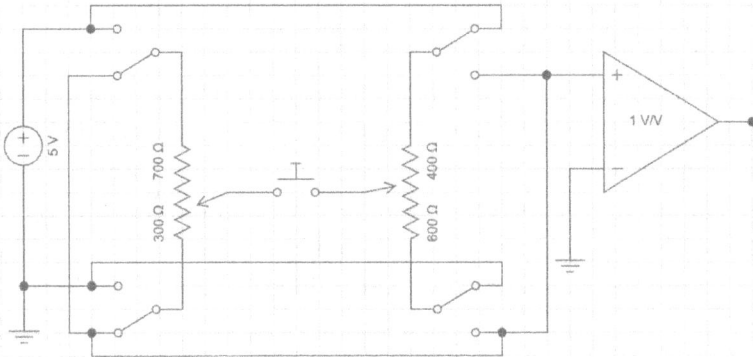

For example, suppose the left potentiometer represented the position of the X-axis, and the right represented the touch position of the Y-axis. A set of switches would allow the ends of the X-axis potentiometer to be connected together and the voltage from these switches would represent the value of the Y-axis position. To get the X-axis position, just reverse this process.

The entire process can be automated with integrated circuits and produce fresh X- and Y-coordinates every millisecond or so.

A screen-sized tablet of this type was put on an Alto CRT, and functioned perfectly. While the resolution was less than that of the high-end digitizing tablet, the prototype tablet only cost a few dollars to build. In the spirit of Zero Mass Design, the result was perfect.

As well as this prototype worked, Xerox chose not to pursue the idea, and when I left the company in 1981, I was given permission to pursue the invention on my own.

A year or so later I was asked by Steve Jobs to design a mouse for Apple, since he had seen one during his trip to Xerox PARC. I told him that while "people had drawn with rocks in the past, now they could draw with a stick!" While my humorous approach to suggesting the tablet as a pointing device may not have worked, he did want to see my tablet working on the Apple II. After the demo, he said he still wanted a mouse, and I recommended that he contact the designers Hovey and Kelley (which later morphed into the firm IDEO). My tablet went back to a shelf in my garage.

A few months later, I heard from a former PARC colleague, George White. George and I had been friends since 1971, and he had left PARC a few years before me to start a speech recognition company, Auricle. After selling Auricle for $2 million, he called and told me he wanted to start another company, perhaps to make computer mice. I told him I had a better idea and to stay home until I got there.

I went to the garage and put an Apple II and my prototype tablet in the car and drove to his place so he could see this invention first-hand. Based on the demo, he put his mouse plans on hold and by the next day he filed the paperwork to start a company. After meeting with venture capitalists, we raised $16 million to get the company going. The Koala name was chosen for the company since we thought children would be the primary users and the Koala was seen as a cuddly mascot.

Early on, in 1983 I filed for a patent (4,484,026) that was issued in 1984. In the meantime products were manufactured for the Apple II, and for the Commodore 64, Atari computers, and the IBM PC. When the Apple Macintosh was released the following year, a KoalaPad was designed for that product as well.

Early-on we did a demo of the KoalaPad at a local computer store called the Byte Shop in Mountain View, California. Everything worked perfectly, but some people were skeptical, asking, for example, "Why would people want to draw pictures on a computer?" While we wouldn't ask this question today, in 1983 personal computers were still largely seen as number-crunchers, not as creative tools for the artistically inclined. This was especially true for traditional customers of the Byte Shop who originally shopped for computer kits, rather than fully assembled and packaged devices.

As time went on, we grew and at one point had about 200 people working for us. The fact that my invention was generating income for

200 people was an item of some pride for George and me.

A few years later the company was shut down, but the resistive touch tablet is still used today in signature machines for credit card transactions.

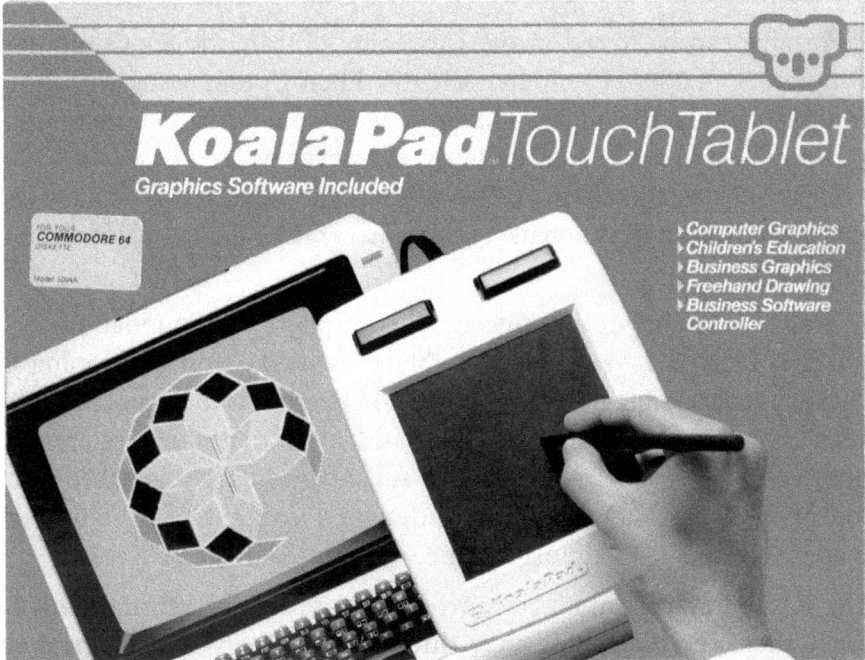

The KoalaPad box, photo courtesy of Giacomo M. Vernoni, oldcomputr.com

A Computer for Everyone

The Raspberry Pi is a credit card-sized computer designed and manufactured in the UK with the initial intention of providing a cheap computing device for education. Since its release, however, it has grown far beyond the sphere of academia.

Its origins can be found in the University of Cambridge's Computer Laboratory in 2006. Computer scientist Eben Upton, along with Rob Mullins, Jack Lang and Alan Mycroft, were concerned that incoming computing undergraduate students had never written a line of code prior to starting their degree. This was largely due to school syllabuses that placed an emphasis on using computers rather than understanding them.

In response to this initial concern, the Raspberry Pi Foundation was formed. Over the next six years the team worked on developing a cheap and accessible device that would help schools to teach concepts such as programming, thus bringing students closer to understanding how computing works. By taking advantage of free software and inexpensive parts, the radically redesigned concept of a computer made it affordable to almost anyone who wanted one.

The Raspberry Pi's initial commercial release was in February 2012. Since then, the board has gone through a number of revisions and is available in several models with a price of $5 to $35 based on the capabilities of the board.

By mid-2021, more than 40 million of these computers had been sold. With the onset of the Covid pandemic, sales have gone through the roof, and some models are very hard to find.

The Raspberry Pi 2 is significantly more powerful than previous versions, opening up many new possibilities and providing capabilities one might find on computers casting ten times as much..

The default operating system chosen for this board is Linux, the free operating system used in an overwhelming number of smartphones, tablets, Chromebooks, and all Apple computers.

Raspberry Pi model 2 picture (CC BY-SA 2.0)

The reason these boards are so inexpensive is because they make use of accessories (HDMI video screens, USB keyboards and mice), users might already have. Programming tools available for this computer range from Scratch, developed for kids at MIT's Lifelong Kindergarten group, to Python and beyond.

Because the Raspberry Pi has numerous pins available for input and output, people have added capabilities (like video cameras) to this device. It's small size has allowed it to be installed in commercial products like 3D printers. While the motivation was to provide a system on which kids could learn to code, its uses have expanded tremendously!

Spring Reverb

Reverberation in acoustics, is a persistence of sound, or echo after a sound is produced. Reverberation is created naturally when a sound or signal is reflected causing numerous reflections to build up and then decay as the sound is absorbed by the surfaces of objects in the space—which could include furniture, people, and air. This is most noticeable when the sound source stops but the reflections continue, their amplitude decreasing, until zero is reached.

Several techniques have been developed to simulate reverberation, and one of these, spring reverberation, is an example of Zero Mass Design.

Spring reverbs use a set of springs (typically two or three) mounted inside a box. They work with a transducer and pickup placed at either end of a spring. They were popular in the 1960s, and were first used by the Hammond company to add reverb to Hammond organs. When a sound signal is inserted at one end of the spring, it is transformed by the spring vibration to produce a delay and echo and then sent to

Two-spring reverb unit like that used in Fender guitar amplifiers.
(Photo from author's amplifier)

a pickup at the other end to be amplified and mixed with the original signal to produce a richer sound. The user has independent control on the volume of the reverberation.

Spring reverbs became popular with guitarists, including the Beatles and other '60's rock bands as they could easily be built into guitar amplifiers. I have one in a Fender guitar amplifier designed in the mid-60's, and still made today.

The inventor of the Hammond Organ, Laurens Hammond was granted a patent on the spring-based mechanical reverberation system in 1941 (2,230,836).

While not producing effects identical to natural reverberation, spring reverb units remain popular today. The utter simplicity of this invention justifies its incorporation in this book. By cleverly using the natural property of a spring, this simple invention produces wonderful sounds using very inexpensive parts.

Traditional Japanese Housing

A traditional Japanese house does not have a designated use for each room aside from the entrance area, kitchen, bathroom, and toilet. Any room can be a living room, dining room, study, or bedroom. This is possible because all the necessary furniture is portable, being stored in *oshiire*, a small section of the house (like a large closet) used for storage. It is important to note that in Japan, the living room is expressed as *ima*, living "space." This is because the size of a room can be changed by altering the partitioning. Large traditional houses often have only one ima (living room/space) under the roof, while kitchen, bathroom, and toilet are attached on the side of the house as extensions.

Partial view of traditional Japanese house (Image credit: Hassan & Mariko at flickr.com/ photos/8451579@N02/3817053452. Licensed cc-by-2.0.)

Somewhat similar to modern Japanese offices, partitions within traditional Japanese houses are created by a variety of movable panels. One of the most common types is *fusuma*, sliding doors made from wood and paper, which are portable and easily removed. Fusuma seal each partition from top to bottom so it can create a mini room within the house. On the edge of a house are *rōka*, wooden floored passages that are similar to hallways. Rōka and ima are partitioned by *shōji*, sliding and portable doors that are also made from paper and wood. Unlike fusuma, paper used for shōji is very thin so outside light can pass through into the house. This was before glass began to be used for sliding doors. Rōka and outside of the house are either partitioned by walls or portable wooden boards that are used to seal the house at night. Extended roofs protect the rōka from getting wet when it rains, except during typhoon season where the house gets sealed completely. Roofs of traditional houses in Japan are made of wood and clay, with tiles or thatched areas on top. The flooring may be made with *tatami* mats, a type of mat used as a flooring material in traditional Japanese-style rooms. Tatamis are made in standard sizes, twice as long as wide, about 0.9 m by 1.8 m depending on the region.

Tatami are covered with a weft-faced weave of soft rush on a warp of hemp or weaker cotton. There are four warps per weft shed, two at each end (or sometimes two per shed, one at each end, to cut costs). The *doko* (core) is traditionally made from sewn-together rice straw, but contemporary tatami sometimes have compressed wood chip boards in their cores. The long sides are usually edged with brocade or plain cloth, although some tatami have no edging.

The beauty of this design, from the standpoint of ZMD, is that (except for the kitchen and bath areas) the layout of the design is completely flexible in that the light weight floor mats and paper-covered walls can be moved.

3D Printing and Mathematics

Makerspaces are found in many schools and libraries, and 3D printers are common elements in many of them. The most common printers use plastic filament which is heated and extruded through a tiny hole onto the item being built while the head and build platform move. If you don't have a 3D printer yet, think of a motorized glue gun to get a sense of how it works. The most popular filament these days is made from PLA plastic because it is safer to use than the other common choice, ABS.

Unlike other 3D fabrication tools like laser cutters, 3D printers produce little to no scrap.

The printer builds objects layer by layer using instructions contained in a 3D imaging file called an STL file (for stereolithography) and these files are created using various design tools including Tinkercad (tinkercad.com) and BlocksCAD (blockscad3d.com). The resultant objects can range from the very simple to the extremely complex. This makes 3D printers perfect tools to make objects that reflect the principles of Zero Mass Design.

Oloid

First, we'll explore the oloid, a three-dimensional curved geometric object that was discovered by Paul Schatz in 1929. It is the convex hull of a skeletal frame made by placing two linked congruent circles in perpendicular planes, so that the center of each circle lies on the edge of the other circle.

The distance between the circle centers equals the radius of the circles. One third of each circle's perimeter lies inside the convex hull, so the same shape may be also formed as the convex hull of the two remaining circular arcs each spanning an angle of $4\pi/3$.

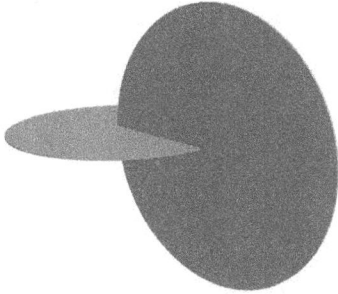

Two intersecting circles

Resulting hull producing an oloid

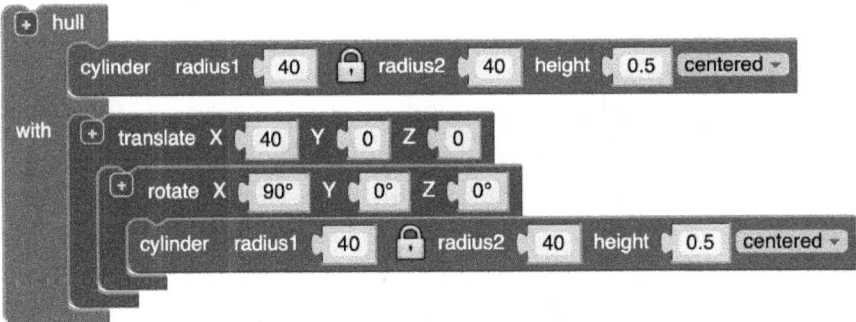

BlocksCAD code for final shape

Note that for the circles I used cylinders with a thickness of 0.5 mm. This makes a smoother print while retaining the desired shape.

The nice thing about BlocksCAD is that the program is easy to read. This is one of the reasons it is my favorite tool for designing 3D objects.

Photo of 3D printed oloid

The resultant shape meets (in my view) the concept of "happy to hold" expressed by the New Bauhaus leader, László Moholy-Nagy. In any case it is pleasing to the eye, as is the Bauhaus-inspired bar of Dove soap.

Reuleaux Triangle

The next shape we'll examine is the Reuleaux triangle. A Reuleaux triangle is a curved triangle with constant width, the simplest and best known curve of constant width other than the circle. It is formed from the intersection of three circular disks, each having its center on the boundary of the other two. Constant width means that the separation of every two parallel supporting lines is the same, independent of their orientation. Because all its diameters are the same, the Reuleaux triangle is one answer to the question "Other than a circle, what shape can a manhole cover be made so that it cannot fall down through the hole?"

This shape is named after Franz Reuleaux, a 19th-century German engineer who pioneered the study of machines for translating one type of motion into another, and who used Reuleaux triangles in his designs. However, these shapes were known before his time, for instance by the designers of Gothic church windows, by Leonardo da Vinci, who used it for a map projection, and by Leonhard Euler in his study of constant-width shapes. Other applications of the Reuleaux triangle include giving the shape to guitar picks, fire hydrant nuts, pencils, as well as in graphic design in the shapes of some signs and corporate logos.

Among constant-width shapes with a given width, the Reuleaux triangle has the minimum area and the sharpest (smallest) possible angle (120°) at its corners. By several numerical measures it is the farthest from being centrally symmetric. It provides the largest constant-width shape avoiding the points of an integer lattice, and is closely related to the shape of the quadrilateral maximizing the ratio of perimeter to diameter. It can perform a complete rotation within a square while at all times touching all four sides of the square, and has the smallest possible area of shapes with this property. However, although it covers most of the square in this rotation process, it fails to cover a small fraction of the square's area, near its corners. Because of this property of rotating within a square, the Reuleaux triangle is also sometimes known as the Reuleaux rotor.

A variation in this shape is used as the rotor in a Wankel engine, is a type of internal combustion engine using an eccentric rotary design to convert pressure into rotating motion.

Compared to the reciprocating piston engine, the Wankel engine has more uniform torque; less vibration; and, for a given power, is more compact and weighs less.

The rotor, which creates the turning motion, is similar in shape to a Reuleaux triangle, except the sides have slightly less curvature. Wankel engines deliver three power pulses per revolution of the rotor using the Otto cycle. However, the output shaft uses toothed gearing to turn three times faster giving one power pulse per revolution.

Wankel engine showing rotor
(Softeis at German Wikipedia, CC BY-SA 3.0, via Wikimedia Commons)

This kind of engine was used in the Mazda RX-7 car, one of which I used for years. When the engine was running, you could scarcely feel the vibrations. Unfortunately, there was a slight leakage in the rotor assembly meaning that the car had a hard time meeting current emission requirements.

Fortunately, the Reuleaux triangle is easy to print. The BlocksCAD code is shown below:

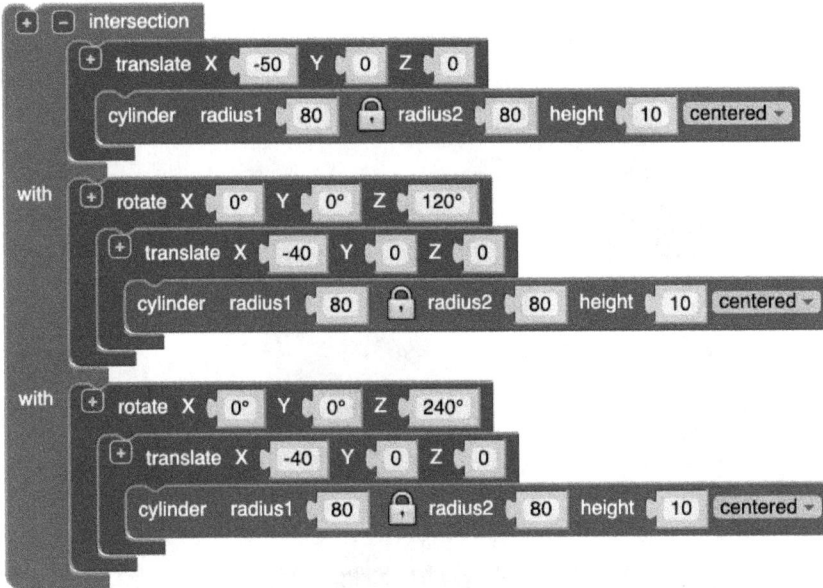

This shape is made from the intersection of three circular disks with a radius of 80 mm, each rotated from the other by 120° and offset from the center by 40 mm.

3D printed Reuleaux triangle

Another characteristic of this shape is that it fits perfectly in a square frame.

Reuleaux triangle in a square frame

As a result it is possible to make a drill bit that makes almost square holes (Smith, 2011).

Möbius Strip

The final shape we'll examine is the Möbius strip. This shape can be made with a strip of paper whose ends are glued together after giving the strip a half turn. Children can observe that this shape only has one side.

Instead of paper, we can make the shape on a 3D printer after generating the shape with a popular math exploration tool, GeoGebra. The version we'll use allows the creation of 3D printable files (geogebra.org).

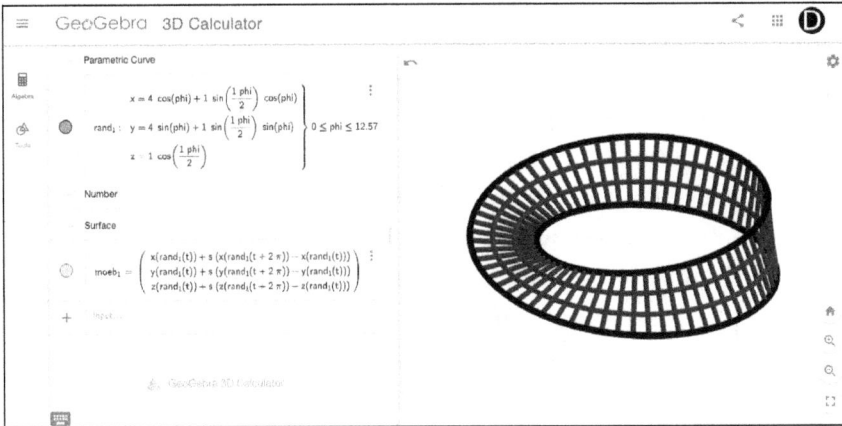

Möbius strip shown in Geogebra with part of the parametric equation that generates the shape

3D printed Möbius strip

While the programming for this shape is somewhat complex, the result is quite simple.

The shapes explored in this chapter fit perfectly with the principles of Zero Mass Design.

IKEA Furniture

The IKEA corporation was started in Sweden in 1943 by 17-year-old Ingvar Kamprad, and IKEA has been the world's largest furniture retailer since 2008. Their designs are quite modern, and by having the furniture packed in easy-to assemble kits, the prices are minimal. Eco-friendly materials and packaging is used extensively, and their designs are quite elegant, very much in the spirit of the Bauhaus.

For example, the Poäng chair, designed by Noboru Nakamura uses a bent plywood frame, reminiscent of Charles and Ray Eames, the pioneers of modern furniture design using bent plywood.

Poäng chair from IKEA

Another elegant design is the Kyrre table which has a top similar to the Reuleaux triangle explored in the chapter on 3D printing.

Kyrre stool/table from IKEA

These are but two examples of ZMD in action. If you live close to an IKEA store, you might enjoy going to their showroom and looking at their designs from a designer's perspective. For example, their Underhälla children's block set is somewhat similar to the Bauhaus Bauspiel, a toy designed by Bauhaus apprentice Alma Siedhoff-Buscher in 1923 (Wikipedia, 2021).

Underhälla block set from IKEA

ZMD and the Future of Our Planet

About 66 million years ago, the Earth was struck by an asteroid 10 to 15 km wide which led to a sudden mass extinction of three-quarters of the plant and animal species. With the exception of some ectothermic species such as sea turtles and crocodilians, no tetrapods weighing more than 25 kilograms survived. It marked the end of the Cretaceous period, and with it the Mesozoic Era, while heralding the beginning of the Cenozoic Era, which continues to this day.

This theory was originally proposed in 1980 by a team of scientists led by Luis Alvarez and his son, Walter. The impact hypothesis, also known as the Alvarez hypothesis, was bolstered by the discovery of the 180 km Chicxulub crater in the Gulf of Mexico's Yucatán Peninsula in the early 1990s (Cretaceous–Paleogene Extinction Event).

While the climate change that led to that extinction was caused by an outside event, our current climate change challenges appear to be made by us, and it is our obligation, as stewards of the planet, to do something to change things before the damage is irreversible. The Earth is the only habitable planet we have close by, so it falls to us to do what we can to fix the planet before it is too late.

In addition to our climate challenges, we are in the midst of a Covid pandemic. Research at Harvard University's school of public health suggests that the spread of this deadly disease might be facilitated by climate change (Coronavirus and Climate Change, 2020). In other words, these events may be coupled.

By the end of 2021, more than 5.3 million people died worldwide from infection by the Sars Covid 19 virus, with over 800,000 of those dying in the United States. With new strains joining the fray, this pandemic is a moving target with no clear end in sight (Coronavirus Resource Center, 2020).

Is it that far a stretch to see some parallels to the extinction of the dinosaurs 66 million years ago? The fact is that we must do something to protect the future of humanity for us and subsequent generations.

Buckminster Fuller expressed hope when he said:

> *I do know that technologically humanity now has the opportunity, for the first time in its history, to operate our planet in such a manner as to support and accommodate all humanity at a substantially more advanced standard of living than any humans have ever experienced. This is possible not because we have found more physical resources. We have always had enough resources. What has happened that now makes the difference is that we have vastly increased our know-how of specialized innovations, all of which invisible realization integrates to make possible success for all. (Fuller, 2008, 11)*

It is my view that some of the principles of Zero Mass Design might shed light on things we can do to reduce the levels of climate change and keep our planet habitable for generations to come.

Reducing our dependence on fossil fuels for power generation is one step, and the Canadian province of Ontario (home to Canada's largest city, Toronto) provides a fine example of what can be done.

In 2018, Ontario generated 151 terawatt hours (TW.h) of electricity, which is approximately 23% of total Canadian generation. Ontario is the 2nd largest producer of electricity in Canada and has a generating capacity of 40 671 megawatts (MW).

In 2018, about 96% of electricity in Ontario was produced from zero-carbon emitting sources: 60% from nuclear, 26% from hydroelectricity, 7% from wind, and 2% from solar. The remainder is primarily from natural gas, with some biomass.

Australia is making huge strides in wind power, and could reach 100% renewable energy in a few years (Parkinson, 2021).

Because greenhouse gas production has been reduced to only 4%, Ontario is a leader in green energy. On the automotive front, the high price of gasoline (compared with the United States) has driven the growth in hybrid and electric cars. On the plug-in electric car front, the growth rate is staggering. In 2018:

- 93,091 electric vehicles (EV) were on the road in Canada
- The number of EVs increased by 90% in the country
- EV sales grew by 125% compared to 2017
- EV sales counted for 2.2% of all passenger vehicle sales

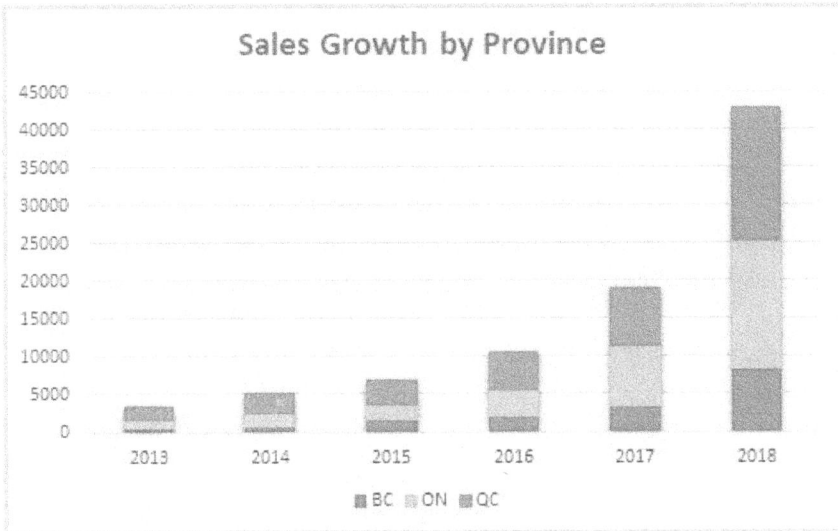

Sales growth of electric vehicles for British Columbia, Ontario, and Quebec. This data is provided by EMC (Electric Vehicle Sales, 2019)

The US state of California has taken a very aggressive position on this topic. In 2021 Governor Gavin Newsom signed a bill requiring that all cars sold in 2035 will have zero carbon emissions, and this law applies to large vehicles (trucks, vans) in 2045. Fifteen other states are considering similar legislation. Even high mileage efficient hybrids will be banned since they use a gas engine to charge the battery used for powering the car electric motor.

In 2018, the total number of plug-in electric vehicles in California was over 160,000, making it a leader in the US, and eclipsing the number of electric cars in Canada.

When it comes to Zero Mass Design, opportunities abound in everything from battery charging stations (e.g., How can a power source be replenished in the time it would take to put gas in a traditional car?), to car design and the design of other vehicles.

But what about the carbon dioxide already in the air? Later this year, Project Vesta (Project Vesta, 2021) plans to spread sand made from a green volcanic mineral known as olivine across beaches. The waves will further break down the highly reactive material, accelerating a series of chemical reactions that pull the greenhouse gas, CO2, out of the air and lock it up in the shells and skeletons of mollusks and corals.

Volcanic rock containing olivine (CC BY-SA 3.0)

This process, along with other forms of what's known as enhanced mineral weathering, could potentially store hundreds of trillions of tons of carbon dioxide. That's far more carbon dioxide than humans have pumped out since the start of the Industrial Revolution. Unlike methods of carbon removal that rely on soil, plants, and trees, it would be effectively permanent. And Project Vesta at least believes it could be cheap, on the order of $10 per ton of stored carbon dioxide once it's done on a large scale.

As Ursula von der Leyen said in her 2021 State of the Union Address as head of the European Union:

Clearly something is on the move. And this is what the European Green Deal is all about. In my speech last year, I announced our target of at least 55% emission reduction by 2030. Since then we have together turned our climate goals into legal obligations. And we are the first major economy to present comprehensive legislation in order to get it done.

The goal is simple. We will put a price on pollution. We will clean the energy we use. We will have smarter cars and cleaner airplanes.

And we will make sure that higher climate ambition comes with more social ambition. This must be a fair green transition.

Under her guidance, The New European Bauhaus was created to do projects related to climate change and seeks participation from around the world. The challenges they face address the global climate crisis that affects all countries.

It is worthwhile to see if there is some silver lining in the pandemic cloud. For example, the self-imposed quarantine has caused many office workers to work from home. This has not only cut down on car exhaust emissions, it has also made it possible for people to spend more time with their families. Face-to-face meetings are being handled through video conferences. All this has a positive effect on the environment. Of course factories, restaurants and traditional stores still need physical workers, but it is amazing to see the absence of traffic jams in dense areas like Silicon Valley in California.

The benefits are so great for some companies that Twitter has decided that its employees could work from home even when the pandemic is over. Twitter's decision to allow its 5,200 primarily San Francisco-based employees to decide where they want to work has major implications for everything from its real estate and salaries to workplace culture. The company could potentially usher in a new model for attracting and retaining talent based on worker-centric values of flexibility, autonomy and satisfaction.

The decisions led by Twitter and followed by other companies could soon herald the end of an era when great ideas at work were born out of daily in-person interactions — a hallmark of Silicon Valley thinking. Apple founder Steve Jobs described it as the creativity that comes from serendipitous run-ins with colleagues, saying that "creativity comes from spontaneous meetings, from random discussions."

Just six months after the coronavirus outbreak led thousands of companies to mandate that their employees work from home, 35 percent of the full-time U.S. labor force is still working remotely, compared to 2 percent prior to the pandemic, according to an August, 2021 survey of over 30,000 professionals by Nicholas Bloom, a Stanford University economics professor and his colleagues who specialize in distributed employment (Barreo et al., 2021). Furthermore, many of these workers plan on working from home after the pandemic is over.

While accelerated by the pandemic, Twitter executives feel that the company's shift to distributed work is ultimately about creating a model that gives employees more autonomy and freedom, which they believe improves morale, retention and productivity.

The company has had its share of challenges. Some Twitter employees are struggling with scheduling across time zones, and executives had to formally cut down on video meetings, which spiked as people went home during the pandemic, to avoid Zoom fatigue, executives and employees said in interviews. They had to rethink their performance review system so it wouldn't be biased against remote workers if people return to the office.

The question is how to recreate "a water cooler effect," a serendipity that lets people build connections that aren't transactional.

Clearly, this change alone sparks the need for creative inventions that promote the widespread growth of new ways of working that are ripe for the application of Zero Mass Design methodologies.

One tool that we use in our house is provided by the Buy Nothing project (buynothingproject.org) where people with unused items (furniture, etc.) can list them on the app and people looking for something can find it for free. Typically, items are tagged with the name of the recipient who then picks up the items and puts them to use. In addition to saving money, this project also keeps items in use instead of recycling them in traditional ways.

There is no limit to the creative inventions that will do their part to heal the planet. When placed against the background of global climate change, there are myriad opportunities to be pursued, many of which can lead to new businesses as well as a cleaner environment. This combined effect was stated most eloquently by Ursula von der Leyen (2020) when she said:

> It is more than cutting emissions. It is about green finance. It is about restoring biodiversity. It is about a new circular economy that creates jobs and prosperity while preserving nature. Many things have to change, so that our planet can remain the same for the next generation.

Conclusion

The goal of this modest book was to introduce a design methodology that facilitates designs of minimum complexity. In addition to describing the techniques of Zero Mass Design, we also showed a few examples from different fields, some of which resulted in commercial products.

The underlying ideas differ significantly from traditional design methodologies, and may take some time getting used to. As mentioned early-on in this book, these principles may not apply to all design projects, and you might to start with some simple challenges first before advancing to large scale projects, like the resistive touch tablet.

As with any designs you create, be sure to document your work and have it dated and witnessed by an outsider who is not one of the inventors, to protect the date of the invention in the case you later want to file for a patent.

On the topic of patents, don't think that your idea is too simple to be patented. This comes up in patent law. A person having ordinary skill in the art, a person skilled in the art, or simply a skilled person is a legal fiction found in many patent laws throughout the world. This fictional person is considered to have the normal skills and knowledge in a particular technical field (an "art"), without being a genius. He or she mainly serves as a reference for determining, or at least evaluating, whether an invention is non-obvious or not (in U.S. patent law), or involves an inventive step or not (in European patent laws). If it would have been obvious for this fictional person to come up with the invention while starting from the prior art, then the particular invention is considered not patentable.

In fact, if your invention is simple and has not been patented yet, then you should be fine. After all, if it was obvious, someone would have already patented it. This doesn't mean that patent examiners don't make mistakes sometimes. For example, several years after my patent was issued for the tapered resistor, a similar patent was issued to someone else. This could have been resolved in court in my favor, but it wasn't worth pursuing. Your patent attorney will best advise you on how to proceed with your own inventions.

My hope is that you find the ideas in this book useful, and fun!

References

Adams, J. (1980). *Conceptual Blockbusting* (2nd ed.). W W Norton & Co, Inc.

Anderson, C. (2014). *Makers: The New Industrial Revolution*. Crown Business.

Barreo, J. M., Bloom, N. A., & Davis, S. J. (2021, June 3). *Why Working from Home Will Stick*. Stanford Graduate School of Business. Stanford Graduate School of Business. gsb.stanford.edu/faculty-research/working-papers/why-working-home-will-stick

Bosch, R. (2013, November 17). Designing for a better world starts at school: Rosan Bosch at TEDxIndianapolis. YouTube. youtube.com/watch?v=q5mpeEa_VZo

Bosch, R. (2018). *Designing for a Better World Starts at School*. Rosan Bosch Studio.

Coronavirus and Climate Change – C-CHANGE | Harvard TH Chan School of Public Health. (2020, July 6). Harvard T.H. Chan School of Public Health. hsph.harvard.edu/c-change/subtopics/coronavirus-and-climate-change/

Coronavirus Resource Center. (2020). Johns Hopkins Coronavirus Resource Center. coronavirus.jhu.edu

Cretaceous–Paleogene extinction event. (n.d.). Wikipedia. en.wikipedia.org/wiki/Cretaceous-Paleogene_extinction_event

Csikszentmihalyi, M. (2008). *Flow: The Psychology of Optimal Experience*. Harper Perennial Modern Classics.

Design Science: A Framework for Change. (1970). The Buckminster Fuller Institute. bfi.org/design-science/primer/design-science-framework-change

Electric Vehicle Sales in Canada in 2018. (2019). Electric Mobility Canada. emc-mec.ca/new/electric-vehicle-sales-in-canada-in-2018/

Fuller, R. B. (1969). *Ideas and Integrities: A Spontaneous Autobiographical Disclosure* (R. W. Marks, Ed.). Collier Books.

Fuller, R. B. (2008). *Grunch of Giants*. Design Science.

Fuller, R. B., & Applewhite, E. J. (1982). *Synergetics: Explorations in the Geometry of Thinking*. Macmillan.

Kay, A. (1986). The Dynabook-Past, Present and Future. Proceedings of the ACM Conference on The History of Personal Workstations. doi.org/10.1145/12178.2533809

Kay, A., & Goldberg, A. (1977). *Personal Dynamic Media*. Computer, 10(3), 31-42. doi.org/10.1109/c-m.1977.217672

Lange, A. (2020). *The Design of Childhood: How the Material World Shapes Independent Kids*. Bloomsbury USA.

McCracken, H. (2015, April 29). Maker Faire Founder Dale Dougherty On The Past, Present, And Online Future Of The Maker Movement. Fast Company. fastcompany.com/3045505/maker-faire-founder-dale-dougherty-on-the-past-present-and-online-future-of-the-maker-moveme

McKim, R. H. (1980). *Experiences in Visual Thinking*. Brooks/Cole.

Paley, S. J. (2010). *The Art of Invention: The Creative Process of Discovery and Design*. Prometheus Books.

Papert, S., & Harel, I. (1991). *Situating Constructionism*. MIT Media Lab. web.media.mit.edu/~calla/web_comunidad/Reading-En/situating_constructionism.pdf

Parkinson, G. (2021, December 22). Australia is racing towards 100 per cent renewables. What does that look like? Renew Economy. reneweconomy.com.au/australia-is-racing-towards-100-per-cent-renewables-what-does-that-look-like/

Project Vesta. (2021). Project Vesta. vesta.earth

Rugg, H. O., & Shumaker, A. (1928). *The Child-centered School: An Appraisal of the New Education*. World Book Company.

Saidani, M., & Remise, E. (2002). Light-Weight Self-Stressed Systems Or Tensegrity. Journal Of The International Association For Shell And Spatial Structures, 43(140), 179-185.

Shaffer, D. W. (2008). *How Computer Games Help Children Learn*. Palgrave Macmillan.

Smith, S. (2011, October 3). Drilling Square Holes with a Reuleaux Triangle — DO IT: Projects, Plans, and How-tos. mike senese. mikesenese.com/DOIT/2011/10/drilling-square-holes-with-a-reuleaux-triangle/

Sokolov, M. (2018). *Practicing Mindfulness*. Althea Press.

Thornburg, D. (2014). *From the Campfire to the Holodeck: Creating Engaging and Powerful 21st Century Learning Environments*. Wiley.

Turing, A. J. (1950, October). Computing Machinery and Intelligence. Mind, LIX(236), 433-460. doi.org/10.1093/mind/LIX.236.433

von der Leyen, U. (2020, December 12). Speech by the President at the Climate Ambition Summit. European Commission. ec.europa.eu/commission/presscorner/detail/en/speech_20_2403

Wikipedia. (2021). Alma Siedhoff-Buscher. Wikipedia. en.wikipedia.org/wiki/Alma_Siedhoff-Buscher

Also From Constructing Modern Knowledge Press

Constructing Modern Knowledge Press publishes books for modern learning. See more at cmkpress.com.

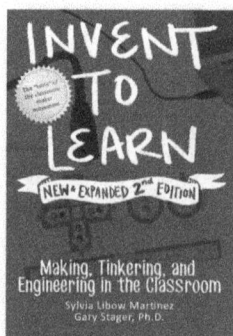

Invent to Learn: Making, Tinkering, and Engineering in the Classroom

by Sylvia Libow Martinez and Gary S. Stager
An all new and expanded edition of the book called "the bible of the Maker Movement in classrooms," *Invent To Learn* has become the most popular book for educators seeking to understand how modern tools and technology can revolutionize education.

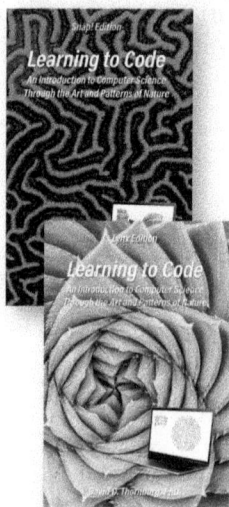

Learning to Code – An Invitation to Computer Science Through the Art and Patterns of Nature (Snap! or Lynx Editions)

by David Thornburg

These are books about discovery—the discoveries each of us can make when finding beauty in geometric patterns, beauty in mathematics, and beauty in computer programming. This is also a guide for teaching children to program computers in uniquely powerful ways.

Available in either a Lynx or Snap! edition—two powerful programming languages designed for learning.

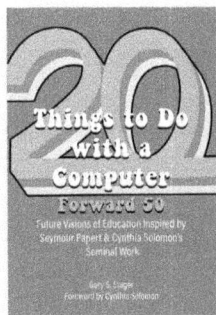

20 Things to Do with a Computer *Forward 50*: Future Visions of Education Inspired by Seymour Papert and Cynthia Solomon's Seminal Work

edited by Gary S. Stager, foreword by Cynthia Solomon
In 1971, Cynthia Solomon and Seymour Papert published *Twenty Things to Do with a Computer*, a revolutionary document that would set the course of education for the next fifty years and beyond. This book is a celebration of the vision set forth by Papert and Solomon a half-century ago. Four dozen experts from around the world invite us to consider the original provocations, reflect on their implementation, and chart a course for the future through personal recollections, learning stories, and imaginative scenarios.

The Art of Digital Fabrication: STEAM Projects for the Makerspace and Art Studio

By Erin E. Riley
Integrate STEAM in your school through arts-based maker projects using digital fabrication tools commonly found in makerspaces like 3D printers, laser cutters, vinyl cutters, and CNC machines. Full color pages showcase the artistic and technical work of students that results from combining art with engineering and design. Written by an educator with experience in art and maker education, this volume contains over twenty-five makerspace tested projects, a material and process inventory for digital fabrication, guides for designing with software, and how-tos for using digital fabrication machines.

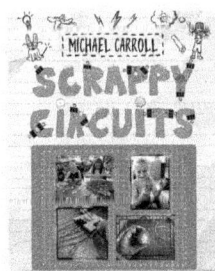

Scrappy Circuits

by Michael Carroll
The best dollar you'll ever spend on a child's STEAM education! Scrappy Circuits is an imaginative "do-it-yourself" way to learn about electrical circuits for less than $1 per person. Raid your junk drawer for simple office supplies, add a little cardboard, pay a visit to a local dollar store, and you are on your way to countless fun projects for learning about electronics.

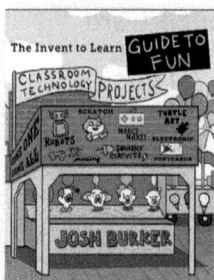

The Invent to Learn Guide to Fun

by Josh Burker

The Invent to Learn Guide to Fun features an assortment of insanely clever classroom-tested maker projects for learners of all ages. Josh Burker kicks classroom learning-by-making up a notch with step-by-step instructions, full-color photos, open-ended challenges, and sample code. Learn to paint with light, make your own Operation Game, sew interactive stuffed creatures, build Rube Goldberg machines, design artbots, produce mathematically generated mosaic tiles, program adventure games, and more!

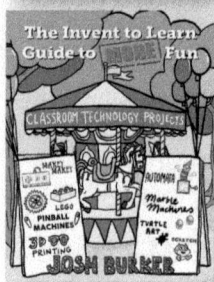

The Invent to Learn Guide to MORE Fun

by Josh Burker

Josh Burker is back with a second volume of all new projects for learners who just want MORE! Insanely clever classroom-tested "maker" projects for learners of all ages with coding, microcontrollers, 3D printing, LEGO machines, and more! The projects feature step-by-step instructions and full-color photos.

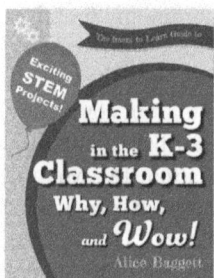

The Invent to Learn Guide to Making in the K-3 Classroom: Why, How, and Wow!

by Alice Baggett

This full color book packed with photos is a practical guide for primary school educators who want to inspire their students to embrace a tinkering mindset so they can invent fantastic contraptions. Veteran teacher Alice Baggett shares her expertise in how to create hands-on learning experiences for young inventors so students experience the thrilling process of making—complete with epic fails and spectacular discoveries.

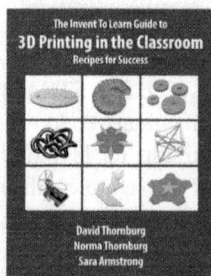

The Invent to Learn Guide to 3D Printing in the Classroom: Recipes for Success

by David Thornburg, Norma Thornburg, and Sara Armstrong

This book is an essential guide for educators interested in bringing the amazing world of 3D printing to their classrooms. Eighteen fun and challenging projects explore science, technology, engineering, and mathematics, along with forays into the visual arts and design.

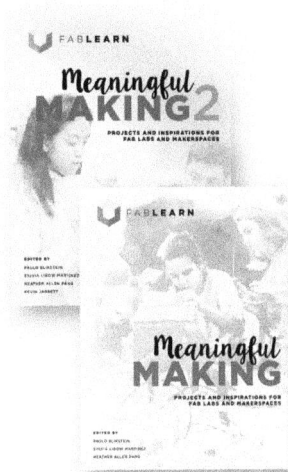

Meaningful Making: Projects and Inspirations for Fab Labs & Makerspaces (Volumes 1 & 2)

Edited by Paulo Blikstein, Sylvia Libow Martinez, Heather Allen Pang

Project ideas, articles, best practices, and assessment strategies from educators at the forefront of making and hands-on, minds-on education. In these two volumes, FabLearn Fellows share inspirational ideas from their learning spaces, assessment strategies and recommended projects across a broad range of age levels. Illustrated with color photos of real student work, the Fellows take you on a tour of the future of learning, where children make sense of the world by making things that matter to them and their communities.

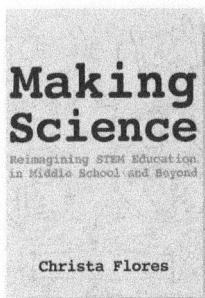

Making Science: Reimagining STEM Education in Middle School and Beyond

by Christa Flores

Anthropologist turned science and making teacher Christa Flores shares her classroom tested lessons and resources for learning by making and design in the middle grades and beyond. Richly illustrated with examples of student work, this book offers project ideas, connections to the Next Generation Science Standards, assessment strategies, and practical tips for educators.

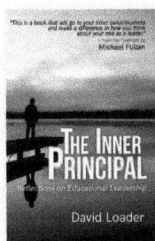

The Inner Principal: Reflections on Educational Leadership

by David Loader

"This is a book that will go to your inner consciousness and make a difference in how you think about your own role as leader." – from the foreword by Michael Fullan

Education Outrage

by Roger C. Schank

Roger Schank has had it with the stupid, lazy, greedy, cynical, and uninformed forces setting outrageous education policy, wrecking childhood, and preparing students for a world that will never exist. The short essays in this book will make you mad, sad, argue with your friends, and take action.